Data Clustering

Edited by Niansheng Tang

Published in London, United Kingdom

Data Clustering
http://dx.doi.org/10.5772/intechopen.95124
Edited by Niansheng Tang

Contributors
Leonard Mushunje, Chiedza Elvina Mashiri, Maxwell Mashasha, Edina Chandiwana, Erick Giovani Sperandio Nascimento, Raphael Souza de Oliveira, Theodore Chadjipadelis, Georgia Panagiotidou, Raja Kishor Duggirala, R.S.M. Lakshmi Patibandla, Veeranjaneyulu N, Ye-Mao Xia, Jian-Wei Gou, Qi-Hang Zhu, Niansheng Tang, Ying Wu

First published in London, United Kingdom, 2022 by IntechOpen
IntechOpen is the global imprint of INTECHOPEN LIMITED, registered in England and Wales, registration number: 11086078, 5 Princes Gate Court, London, SW7 2QJ, United Kingdom
Printed in Croatia

British Library Cataloguing-in-Publication Data
A catalogue record for this book is available from the British Library

Additional hard and PDF copies can be obtained from orders@intechopen.com

Data Clustering
Edited by Niansheng Tang
p. cm.

This title is part of the Artificial Intelligence Book Series, Volume 10
Topic: Machine Learning and Data Mining
Series Editor: Andries Engelbrecht
Topic Editor: Marco Antonio Aceves Fernandez

Print ISBN 978-1-83969-887-3
Online ISBN 978-1-83969-888-0
eBook (PDF) ISBN 978-1-83969-889-7
ISSN 2633-1403

For EU product safety concerns:
IN TECH d.o.o., Prolaz Marije Krucifikse Kozulić 3, 51000 Rijeka, Croatia,
info@intechopen.com or visit our website at intechopen.com.

IntechOpen Book Series

Artificial Intelligence

Volume 10

Aims and Scope of the Series

Artificial Intelligence (AI) is a rapidly developing multidisciplinary research area that aims to solve increasingly complex problems. In today's highly integrated world, AI promises to become a robust and powerful means for obtaining solutions to previously unsolvable problems. This Series is intended for researchers and students alike interested in this fascinating field and its many applications.

Meet the Series Editor

Andries Engelbrecht received the Masters and Ph.D. degrees in Computer Science from the University of Stellenbosch, South Africa, in 1994 and 1999 respectively. He is currently appointed as the Voigt Chair in Data Science in the Department of Industrial Engineering, with a joint appointment as Professor in the Computer Science Division, Stellenbosch University. Prior to his appointment at Stellenbosch University, he has been at the University of Pretoria, Department of Computer Science (1998-2018), where he was appointed as South Africa Research Chair in Artifical Intelligence (2007-2018), the head of the Department of Computer Science (2008-2017), and Director of the Institute for Big Data and Data Science (2017-2018). In addition to a number of research articles, he has written two books, Computational Intelligence: An Introduction and Fundamentals of Computational Swarm Intelligence.

Meet the Volume Editor

Niansheng Tang is a Professor of Statistics and Dean of the School of Mathematics and Statistics, Yunnan University, China. He was elected a Yangtze River Scholars Distinguished Professor in 2013, a member of the International Statistical Institute (ISI) in 2016, a member of the board of the International Chinese Statistical Association (ICSA) in 2018, and a fellow of the Institute of Mathematical Statistics (IMS) in 2021. He received the ICSA Outstanding Service Award in 2018 and the National Science Foundation for Distinguished Young Scholars of China in 2012. He serves as a member of the editorial board of *Statistics and Its Interface* and *Journal of Systems Science and Complexity*. He is also a field editor for *Communications in Mathematics and Statistics*. His research interests include biostatistics, empirical likelihood, missing data analysis, variable selection, high-dimensional data analysis, Bayesian statistics, and data science. He has published more than 190 research papers and authored five books.

Contents

Preface

Clustering is the process of grouping or classifying data points into several different groups or classes where several similar data points are organized into the same cluster or group according to some criterion such as similar features or characteristics. That is, a cluster or group is a collection of several or many data points that have some similar features or characteristics and have different features or characteristics from data points in other clusters or groups. Clustering can be used to obtain the distribution of data, the features of each cluster or group, and analyze some special clusters or groups. Data clustering has many important practical applications in exploratory pattern analysis, grouping, and decision-making, such as classifying sales data to reflect consumer buying behavior or classifying network data to explore communication patterns or grouping student data to reveal their sex characteristics or some field's specialty or special skill or knowledge. The similarity between data points plays an important role in clustering. To this end, many statistics have been developed to measure similarity, for example, Mahalanobis distance, K-means, K medoids, Wasserstein distance, Kullback-Leibler divergence, and others. In addition, many algorithms have been developed for data clustering, for example, partitional clustering, hierarchical clustering, density-based clustering, and model-based clustering. However, these existing methods cannot directly be applied to more complicated data clustering such as nonlinear separable patterns, heterogeneity data, jump-diffusion models, and Brazilian legal documents for natural language processing. As such, this book introduces some novel approaches to deal with these complicated data or models. Also, to stimulate readers' interest, this book introduces the application of data clustering methods developed recently for insurance, psychology, and pattern recognition, and survey data.

This book includes three sections and seven chapters.

Section I includes one chapter that discusses the development of data clustering, including measures of similarity or dissimilarity for data clustering, data clustering algorithms and assessment of clustering algorithms.

Section II introduces clustering methods and includes three chapters. In Chapter 2, Dr. Lakshmi, and Dr. Veeranjaneyulu scientifically review the widely used clustering algorithm. In Chapter 3, Professor Nascimento and Dr. Souza de Oliveira introduce a data clustering method based on the similarity of Brazilian legal documents using natural language processing approaches. In Chapter 4, Dr. Xia, Dr. Zhu, and Dr. Gou present a Bayesian model-based clustering technique for assessing the heterogeneity of a two-part model and investigate its application to cocaine use data.

Section III includes three chapters that focus on the application of recently developed clustering methods. In Chapter 5, B.Sc. Mushunje, Mrs. Mashiri, Dr. Chandiwana, and Mr. Mashasha discuss the application of jump-diffusion models to insurance claim estimation. In Chapter 6, Dr. Duggirala studies fuzzy perceptron learning for non-linearly separable patterns. Finally, in Chapter 7, Professor Chadjipadelis and Dr. Panagiotidou present a semantic map including bringing together groups and discourses.

I was invited to edit this book after the publication of *Bayesian Analysis for Hidden Markov Factor Analysis Models*, which I co-wrote with Yemao Xia, Xiaoqian Zeng, and my previously edited book, *Bayesian Inference on Complicated Data*. I am very grateful to Dr. Maja Bozicevic for his kind invitation to edit this book and for providing me the chance to work with my aforementioned coauthors. I would also like to thank all the chapter authors for their contributions. I hope this book will be of great interest to statisticians, engineers, decision-makers, data analysts, biologists, ecologists, and AI and machine learning researchers.

Niansheng Tang
Department of Statistics,
Yunnan University,
Kunming, China

Section 1

Introduction

Chapter 1

Introductory Chapter: Development of Data Clustering

Niansheng Tang and Ying Wu

1. Introduction

Data clustering is a popular method in statistics and machine learning and is widely used to make decisions and predictions in various fields such as life science (e.g., biology, botany, zoology), medical sciences (e.g., psychiatry, pathology), behavioral and social sciences (e.g., psychology, sociology, education), earth sciences (e.g., geology, geography), engineering sciences (e.g, pattern recognition, artificial intelligence, cybernetics, electrical engineering), and information and decision sciences (e.g., information retrieval, political science, economics, marketing research, operational research) [1]. Clustering analysis aims to group individuals into a number of classes or clusters using some measure such that the individuals within classes or clusters are similar in some characteristics, and the individuals in different classes or clusters are quite distinct in some features.

2. Measures of similarity or dissimilarity

There are a lot of measures of similarity or dissimilarity for data clustering. Generally, assessing the similarity of individuals in terms of the number of characteristics, which can be regarded as the points in space (e.g., a plane, the surface of a sphere, three-dimensional space, or higher-dimensional space) directly relates to the concept of distance from a geometrical viewpoint [1]. The widely used measures include Euclidean distance, Manhattan distance (also called city-block distance), and Mahalanobis distance for measuring the similarity of two data points. Euclidean distance depends on the rectangular coordinate system, Manhattan distance depends on the rotation of the coordinate system, but Euclidean and Manhattan distances do not consider the correlation between data variables and data dimensions. Mahalanobis distance can be regarded as a correction of Euclidean distance, the dependence of the data points is described by covariance matrix, which can be used to deal with the problem of non-independent and identically distributed data. In addition, there are other distances such as chebychev distance, power distance, and sup distance.

In many applications, different types of data are related to different distances. For example, the simple matching distance is used to measure the similarity of two categorical data points; a general similarity coefficient is adopted to measure the distance of two mixed-type data points or the means of two clusters; probabilistic model, landmark models, and time series transformation distance are used to measure the similarity of two time-series data points. In particular, Wu et al. [2] considered spectral clustering for high-dimensional data exploiting sparse representation vectors, Kalra et al. [3] presented online variational learning for medical image data clustering, Prasad et al. [4] discussed leveraging variational autoencoders for image clustering, Soleymani et al. [5] proposed a deep variational clustering framework for self-labeling of large-scale medical image data.

Although the aforementioned similarity and dissimilarity measures can be applied to various types of data, other types of similarity and dissimilarity measures such as cosine similarity measure and a link-based similarity measure have also been developed for specific types of data. Also, one may require computing the distance between an individual and a cluster, and the distance between two clusters based on various central data points or representative data points. In these cases, the widely used distances include the mean-based distance, the nearest neighbor distance, the farthest neighbor distance, the average neighbor distance, which are extensions of data point distances. Particularly, the Lance-Williams formula can be used to compute the distances between the old clusters and a new cluster formed by two clusters. Again, to assess the similarity among probability density distributions of random variables, one can use Kullback-Leibler (K-L) distance (relative entropy) and Wasserstein distance. K-L distance does not satisfy three properties of the distance and is asymmetric, while Wasserstein distance possesses three properties of the distance and is symmetric. More importantly, Wasserstein distance can be used to deal with the mixture of discrete and continuous data. To this end, data clustering based on the Wasserstein distance has received a lot of attention over the past years. For example, see [6–9] for dynamic clustering of interval data, complex data clustering, variational clustering, geometric clustering, respectively.

3. Data clustering algorithms

Many useful data clustering algorithms have been developed to cluster individuals into different clusters over the past years. For example, hierarchical clustering algorithm, partitioning algorithm, fuzzy clustering algorithm [10], center-based clustering algorithm, search-based clustering algorithm, graph-based clustering algorithm, grid-based clustering algorithm, density-based clustering algorithm, model-based clustering algorithm, and subspace clustering [11]. Hierarchical clustering algorithm, which divides individuals into a sequence of nested partitions has two key algorithms: agglomerative algorithm and divisive algorithm, and partitioning algorithm are two important clustering algorithms. Fuzzy clustering algorithm allows an individual to belong to two or more clusters with different probabilities, has three major algorithms: fuzzy k-means, fuzzy k-modes, and c-means. Center-based clustering algorithm is more used to cluster large scales and high-dimensional data sets has two major algorithms: k-means and k-modes in which k-means is the most widely used clustering algorithm, and is a non-hierarchical clustering method. Search-based clustering algorithm is usually used to find the globally optimal clustering for fitting the data set in a solution space, its main algorithms include genetic algorithm, tabu search algorithm, and simulated annealing algorithm. Graph-based clustering algorithm is suitable for clustering graphs or hypergraphs via the dissimilarity matrix of the data set. Grid-based clustering algorithm is sequentially implemented by partitioning the data space into a finite number of cells, estimating the cell density for each cell, sorting the cells with their densities, determining cluster centers, and traversal of neighbor cells, it can largely reduce the computational complexity. Density-based clustering algorithm is clustered according to dense regions separated by low-density regions, can be used to cluster any shaped clusters but is not suitable for high-dimensional data sets. The commonly used density-based clustering algorithms include DBSCAN (Density-based spatial clustering of application with noise), which cannot deal with clustering for data sets with different densities, OPTICS (Ordering points to identify the clustering structure) which can solve the clustering problem for data sets with different densities and outliers, BRIDGE, DBCLASD, DENCLUE, and CUBN algorithms. Recently Ma et al. [12] developed a density-based radar scanning clustering algorithm that can discover and accurately extract individual clusters by employing the radar scanning strategy.

Model-based clustering algorithm becomes an increasingly popular tool and is conducted by assuming that data sets under consideration come from a finite mixture of probability distributions, and each component of the mixture represents a different cluster, which indicates that it requires knowing the number of components in the mixture including finite mixture model (a parametric method) and infinite mixture model (a nonparametric method), the clustering kernel including multivariate Gaussian mixture models, the hidden Markov mixture models, Dirichlet mixture models, and non-Gaussian distributions-based mixture models. Also, model-based clustering algorithms can be divided into non-Bayesian and Bayesian methods, its implementation is challenging [13]. Recently Goren and Maitra [14] developed a clustering methodology using the marginal density for the observed values assuming a finite mixture model of multivariate t distributions for partially recorded data. For clustering problems with missing data, the most common treatment is deletion or imputation. Deletion methods may lead to poor clustering performance when the missing data mechanism is not missing completely at random. In contrast, imputation method using a predicted value to impute each missing value may lead to a better clustering performance when the missing data mechanism is missing at random. But it is rather difficult to impute a suitable value for each missing value for missing not at random. The defects of deletion and imputation do not consider the missing data structure. Model-based clustering via the finite mixture of the multivariate Gaussian or t distributions has been applied to many fields, for example, see [15, 16].

Subspace clustering is conducted by identifying different clusters embedded in different subspaces of the high-dimensional data, whose clustering has several difficulties: distinguishing similar data points from dissimilar ones due to the same distance between any two data points, and different clusters lying in different subspaces. In this case, dimension reduction techniques such as principal component analysis or feature selection techniques [17, 18]. It is rather difficult to tell readers which algorithm should be used for some settings considered and how to compare novel ideas with the existing results because of its unsupervised learning process. But Gan, Ma and Wu [11] gave a comprehensive review of the applications of the aforementioned clustering algorithms.

4. Assessment of clustering algorithm

Since data clustering is a non-supervised method, the assessment of clustering algorithm is rather important. In the data clustering, there are no pre-specified clusters, it is rather challenging to find an appropriate index for measuring whether the obtained cluster result is acceptable or not. The process of assessing the results of a clustering algorithm is usually referred to as clustering validity evaluation.

Generally, clustering validity assessment includes judging the quality of clustering results, the degree to which the clustering algorithm is suitable for a special data set, and finding the best number of clusters. There are two criteria for clustering validity, for example, compactness that the individual within each cluster should be as close to each other as possible and the common measure of compactness is the variance, and separation that the clusters themselves should be separated and the commonly used methods for measuring the distance between two different clusters are the distance between the closest individual of the clusters, distance between the most distant individuals and distance between the centers of the clusters. There are three indices for assessing the results of the clustering algorithm, for example, internal indices measuring the inter-cluster validity, external indices measuring the intra-cluster validity, and relative indices.

Both internal and external indices are based on statistical methods and involve intensive computation. The comprehensive review can refer to [19]. With the increase in the dimension of data points and variables, the cluster analysis method needs to be combined with the corresponding dimension reduction technology. Extracting features through dimension reduction technology and using features to realize clustering is a method of cluster analysis of high-dimensional data.

5. Future interesting topics

Some interesting research fields in the future include model-based clustering with missing not at random data and skew-normal or skew-t distribution, model-based tensor clustering, which is a challenging topic due to the correlation structure, ultrahigh-dimension and sparsity of tensor data, and the dimension of each mode of the tensors growing at an exponential rate of the sample size, and high-dimensional and ultrahigh-dimensional data clustering that is also challenging due to sparsity of data. In these cases, data clustering needs to incorporate the dimension reduction technique and imputation technique of missing data. Also, variational and distributed techniques for data clustering may be important and challenging research with the development of computing techniques.

Acknowledgements

We would like to cordially thank Maja Bozicevic for polishing this chapter and fruitful remarks regarding chapter structure. This work was funded by NSFC grant 11731011.

Author details

Niansheng Tang* and Ying Wu
Department of Statistics, Yunnan University, Kunming, P.R. China

*Address all correspondence to: nstang@ynu.edu.cn

References

[1] King RS. Clustering Analysis and Data Mining: An Introduction. Dulles: Mercury Learning and Information; 2015

[2] Wu S, Feng X, Zhou W. Spectral clustering of high-dimensional data exploiting sparse representation vectors. Neurocomputing. 2014;**135**:229-239

[3] Kalra M, Osadebey M, Bouguila N, Pedersen M, Fan W. Online variational learning for medical image data clustering. In: Bouguila N, Fan W, editors. Mixture Models and Applications. Unsupervised and Semi-Supervised Learning. Cham: Springer; 2020

[4] Prasad V, Das D, Bhowmick B. Variational clustering: Leveraging variational autoencoders for image clustering. In: 2020 International Joint Conference on Neural Networks (IJCNN); 19-24 July 2020; Glasgow, UK. Washington, US: IEEE; 2020. pp. 1-10

[5] Soleymain F, Eslami M, Elze T, Bischl B, Rezaei M. Deep variational clustering framework for self-labeling of large-scale medical images. In: Proc. SPIE 12032, Medical Imaging 2022: Image Processing; 4 April 2022; San Diego, California, US. 2022. pp. 68-76. DOI: 10.1117/12.2613331

[6] Irpino A, Verde R. Dynamic clustering of interval data using a Wasserstein-based distance. Pattern Recognition Letters. 2008;**29**:1648-1658

[7] Irpino A. Clustering linear models using Wasserstein distance. In: Palumbo F, Lauro C, Greenacre M, editors. Data Analysis and Classification. Studies in Classification, Data Analysis, and Knowledge Organization. Berlin: Springer; 2009

[8] Mi L, Zhang W, Gu X, Wang Y. Variational Wasserstein clustering. Computer Vis ECCV. 2018;**11219**: 336-352

[9] Mi L, Yu T, Bento J, Zhang W, Li B, Wang Y. Variational Wasserstein Barycenters for geometric clustering. 2020. DOI: 10.48550/arXiv.2002.10543

[10] Abonyi J, Feil B. Cluster Analysis for Data Mining and System Identification. Berlin: Birkhauser Verlag AG; 2007

[11] Gan G, Ma C, Wu J. Data Clustering: Theory, Algorithms, and Applications. Pennsylvania: SIAM; 2007

[12] Ma L, Zhang Y, Leiva V, Liu SZ, Ma TF. A new clustering algorithm based on a radar scanning strategy with applications to machine learning. Expert System with Applications. 2022;**191**:116143

[13] Melnykov V. Challenges in model-based clustering. WIREs Computational Statistics. 2013;**5**:135-148

[14] Goren EM, Maitra R. Fast model-based clustering of partial records. Statistics. 2022;**11**:e416

[15] Lin TI. Learning from incomplete data via parameterized t mixture models through eigenvalue decomposition. Computational Statistics and Data Analysis. 2014;**71**:183-195

[16] Wang WL, Lin T. Robust model-based clustering via mixtures of skew-t distributions with missing information. Advances in Data Analysis and Classification. 2015;**9**:423-445

[17] Yan X, Tang N, Xie J, Ding X, Wang Z. Fused mean-variance filter

for feature screening. Computational Statistics and Data Analysis. 2018;**122**: 18-32

[18] Xie J, Lin Y, Yan X, Tang N. Category-adaptive variable screening for ultrahigh dimensional heterogeneous categorical data. Journal of the American Statistical Association. 2020;**115**:747-760

[19] Lazzerini B, Jain LC, Dumitrescu D. Cluster validity. In: Fuzzy Sets & Their Application to Clustering & Training. Boca Raton: CRC Press; 2020. pp. 479-516

Section 2

Clustering Methods

Chapter 2

Clustering Algorithms: An Exploratory Review

R.S.M. Lakshmi Patibandla and Veeranjaneyulu N

Abstract

A process of similar data items into groups is called data clustering. Partitioning a Data Set into some groups based on the resemblance within a group by using various algorithms. Partition Based algorithms key idea is to split the data points into partitions and each one replicates one cluster. The performance of partition depends on certain objective functions. Evolutionary algorithms are used for the evolution of social aspects and to provide optimum solutions for huge optimization problems. In this paper, a survey of various partitioning and evolutionary algorithms can be implemented on a benchmark dataset and proposed to apply some validation criteria methods such as Root-Mean-Square Standard Deviation, R-square and SSD, etc., on some algorithms like Leader, ISODATA, SGO and PSO, and so on.

Keywords: partition, evolutionary, algorithms, clustering

1. Introduction

Clustering is unique to the utmost essential methods in data mining. Clustering is one of the major tasks of grouping the objects which have more attributes from different classes and the objects that belong to the same class are similar. Clustering is an eminent research field that has been used in various areas like Big Data Analytics, Statistics, Machine Learning, Artificial Intelligence, Data Mining, Deep Learning, and so on. Diverse algorithms have been anticipated for assorted applications in clustering [1]. The evaluation of these algorithms is most essential in unsupervised learning. There are no predefined classes in clustering thus it is complicated to measure suitable metrics. For this, a variety of validation criteria have been implemented [2, 3]. The major disadvantage of these validation criteria is cannot evaluate the arbitrary shaped clusters. As it normally selects a particular point from every cluster and computes the distance of particular points based on some other parameters. Suppose variance is computed based on these parameters.

Data Clustering is appropriated among the dataset dividing into different bunches with the end goal that the examination in the gathering is better than different groups. The dataset is to be apportioned to some degree if the information is similarly conveyed, attempt to distinguish the information of certain groups will fall flat or will prompt acquainted a few segments that are with being fake. Another issue is that the covering of information gatherings. These gatherings are at times diminishing the bunching strategies proficiency. This decline the effectiveness is corresponding to the amount of coverage between the groups. Another issue of bunching calculations is their ability to be created in the method of on the web or disconnected. Web-based grouping is a technique for which an input vector is utilized to reconsider the bunch places according to the situation of the vector. Right now, a process where the focuses of groups are to be presented new information every single time. In disconnected mode, the technique is applied on a preparation informational collection, used to locate the focal point of bunches by examining all the information vectors in the preparation set. The bunch communities are found once they are fixed and used to characterize input vectors later. The systems are introduced right now.

Right now, strategies, transformative techniques for bunching, and group approval criteria are presented in Section 2. The complete investigation of the fundamentals much of the time utilized approval techniques in Section 3. The proposed work has been presented in Section 4.

2. Related work

The issue is to recognize the comparative information things and structure as bunches. There are a few calculations and can be delegated Partitioning bunching, Hierarchical Clustering, Density-based Clustering, and Grid-put together Clustering. Here mostly concentrate concerning Partitioning calculations and developmental calculations on seat mark datasets. Dividing calculations legitimately decays an informational index into a lot of disjoint bunches and to decide various parcels have been utilized sure paradigm capacities. Transformative calculations are gotten from the hard bunching calculations for getting the ideal outcomes. The aftereffects of a bunching calculation are not comparable starting with one then onto the next applied with a few information parameters on the same informational index. To assess the groups some approval measures have been proposed. Smallness and Separation approaches are utilized to quantify the separation between groups. Outside criteria, interior criteria, and relative criteria are the three strategies to assess the consequences of grouping. Outer and inside criteria both can have a high computational interest and are dependent on factual methodologies. The significant downside of these two methodologies is the multifaceted nature of calculations. The relative criteria are the assessment of different groups. Many grouping calculations are executed on more occasions on the same informational index with various information parameters. The fundamental goal of the relative criteria is to choose the best grouping calculation from various outcomes based on approval criteria. These distinctive approval criteria have been actualized [4–9].

2.1 Partitioning methods

These strategies are classified into two different ways, the centroid and medoid calculations. The centroid calculations are the calculations to speak to each bunch with the assistance of the greatness of the focus of the cases [10, 11]. The medoid calculations are the calculations that speak to each group of the examples storage room to the size place. K-implies calculation is the generally utilized centroid calculation [12]. The k-implies calculation isolates the informational index into k subsets as each point in a given subset is nearest to a similar focus. Ordinarily, the k-implies have some helpful properties, for example, handling on enormous informational collections is productive, over and again stops at neighborhood ideal, having circular shape bunches and touchy to clamor. This calculation goes under the bunching technique since it requires the information ahead of time. The fundamental k-implies calculations principle objective is choosing the exact starting centroids. The most as of late utilized calculation for clear-cut traits is k-modes calculation. Both k-means and k-modes calculations permit cases of bunching by utilizing blended characteristics in the k-models calculation. The disentanglement of normal k-implies has been introduced most as of late. This can be utilized on ball and circle formed information groups with no issue and performs definite bunching without pre-deciding the exact group number. Some conventional grouping calculations produce allotments. In a parcel, all examples have a place with just one single bunch. Along these lines, each bunch in a hard grouping is disjoint.

Fluffy-based grouping stretches out the view to relate each example among each bunch through enrollment work. Generally utilized calculation for this is Fuzzy C-implies calculation, which depends on k-implies. Fluffy C-implies calculation is utilized to locate the run-of-the-mill point in each group. It tends to be viewed as the focal point of the bunch and enrollment of each case in the group. Other delicate bunching calculations have been actualized, based on the Expectation–Maximization calculation [13]. This calculation accepts an easygoing probabilistic model with specific parameters that depict the probabilistic cases of that bunch. The arrangement of FM calculation starts with essential speculations for the Mixture Model parameters. These qualities are utilized to assess the probabilities of bunches for each example. This procedure is rehashed to re-gauge the parameters of those probabilities. The drawback of this calculation is computationally progressively costly. Over-fitting is the issue in the previously mentioned strategy. This issue emerges for two reasons. The initial one is a tremendous number of bunches might be exact. The second one is the likelihood dispersions have more parameters. Completely Bayesian methodology is one of the plausible arrangements right now every parameter has a previous likelihood conveyance. ISODATA is one of the generally utilized solos characterization calculations. It is an iterative calculation and like k-implies. ISODATA calculation split and consolidated the bunches for future refinements. The primary contrast between ISODATA and k-implies is ISODATA permits various bunches while the k-implies expect that the groups are known as apriori. Gradual bunching calculation which is utilized on enormous informational indexes is Leader Algorithm. Pioneer is structure-based calculation and structure different bunch relies upon the request for the informational index which is accommodated calculation.

As indicated by Ashish Goel [14], while looking at k-implies, Fuzzy k-means and k-medoids rather than centroid have been utilized in the middle or Partition Around Medoids. In this way, k-implies utilize the centroid for speaking to the bunch not manage the anomalies. That is, an information object with the most noteworthy estimation of information can be conveyed. This technique handles this with the medoids' portrayal of the bunch as an incredible centroid. Rather than centroid, the predominantly set information object of the group on the inside is called a Medoid. Right now, several information objects have favored discretionarily equivalent to medoids for speaking to k number of bunches. And all other leftover information objects are in the group have a medoid which is like that information object. After consummation of all the procedure of information questions, another medoid is presented in the spot of centroid to speak to bunches in a most ideal manner and once more the entire procedure is persistent. All the information objects have limited to the bunches relies upon the most up-to-date medoids. Medoids correct their position consistently for every cycle. This nonstop procedure is till the remaining medoids sit tight for a move. Inevitably, k groups to speak to a lot of information items can be found. Examination of K-Means, Fuzzy K-Means, and K-Medoids are investigated in the accompanying **Table 1**.

On the other hand, several Evolutionary algorithms have been implemented for optimization. Some of the Evolutionary Algorithms have been explained below.

2.2 Evolutionary algorithms

A Genetic Algorithm is a factual advancement approach. The Genetic Algorithm is a notable calculation that is applied to different ideal plan issues. Also, it decides worldwide ideal arrangements by a consistent variable savvy calculation. Differential Evaluation is additionally like Genetic calculation.

Clonal Selection Algorithm is the developmental calculation for the natural resistant framework. There are two components determination and transformation. These two systems are finished by a record of invulnerable properties. Then again, the blast rate is corresponding to the proclivity, and the transformation rate is conversely relative to liking. The connection among lock and key must fit with one another and afterward, the reaction will work.

Particle Swarm Optimization is a transformative bunching calculation and reenacts the properties of running winged creatures. It follows some situations used to take care of the enhancement issues. Right now, the single arrangement is a winged creature in search, call it a Particle. Each Particle is considered as a point in dimensional space. **Figure 1** shows the process flow of the PSO algorithm.

	K-means	Fuzzy K-means	K-medoids
Complexity	O(ikn)	O (I k (n)2)	O (i k (n-k)2)
Efficiency	Comparatively more	Comparatively more than K-Medoids	Comparatively less
Implementation	Easy	Less complicated than K-Medoids and Complicated to K-Means	Complicated
Sensitive to Outliers?	Yes	No	No
The necessity of convex shape	Yes	Not so much	Not so much
Advance specification of no of clusters 'k'	Required	Required	Required
Does initial partition affects result and runtime?	Yes	Yes	Yes
Optimized for	Separated clusters	Separated cluster and categories data	Separated clusters,

Table 1.
K-means, fuzzy K-means, and K-medoids algorithm comparison details.

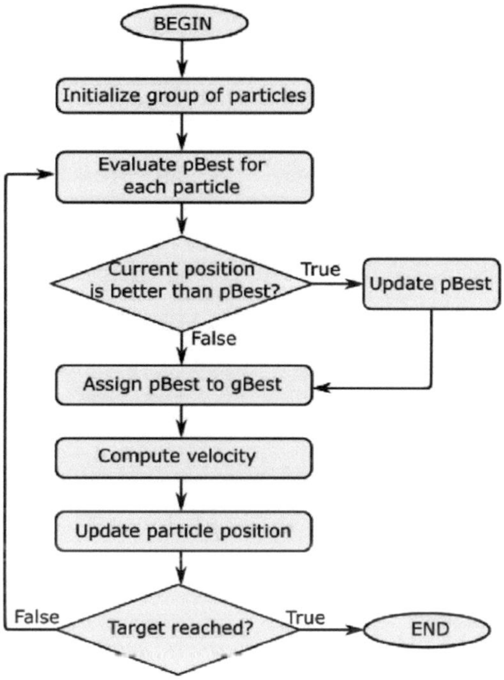

Figure 1.
Flow chart for particle swarm optimization.

Teaching Learning Based Optimization [10] is one of them as of late actualized advancement calculation. In designing applications, it impacts the impact of an instructor on the yield of students in a class is investigated by scientists for taking care of various streamlining issues.

Suresh Satapathy et al. [8] proposed a novel enhancement calculation named Social Group Optimization that relies upon the conduct of people to learn and take care of complex issues. They executed and examine the exhibition of SGO advancement calculation on a few benchmark capacities. Right now, dissected the different human characteristics of life, for example, resilience, fearlessness, dread, and deceitfulness, etc.

Social Group Optimization calculation can be partitioned into two different ways improving stage and securing stage. Every individual's information level in the gathering has been tried and upgraded by the impact of the best one in the gathering in the improving stage. The best individual in the gathering having the information for taking care of issues. Everybody in the gathering improves information with communications to each other in the gathering and best one in the gathering around then.

As per Wen-Jye Shyr [15], to compute and verify the improvement calculations execution estimated two elements of numerical destinations. The exhibitions of these techniques can be depicted for certain perspectives that are demonstrated as follows. The initial one is the ideal point union, which is the key executive for this calculation. The second one is the ideal incentive for exactness. The third one is the absolute number of target calculations. For the most part, there are a lot of issues where assembly speed is dependent on the absolute number of target calculations. The last one is the time taken for the calculation to locate the ideal worth. Even though this is the simplicity of calculation can be unforeseen. Notwithstanding these, a few parameters are made, tried to ensure that the outcomes are set in **Table 2**.

Genetic Algorithm (GA)	Population Size 20 Crossover Probability of 0.6 Mutation Probability 0.005 Iterations 50
Clonal Selection Algorithm (CSA)	Number of Clones Generated 100 Hyper mutation Probability 0.01 Scales of Affinity Proportion Selection 100 Percentage of Random New Cells each Generation 10% Iterations 50
Particle Swarm Optimization (PSO) based Algorithm	Population Size 20 Initial Inertia Weight 0.9 Final Inertia Weight 0.2 Iterations 50

Table 2.
Genetic algorithm, clonal selection algorithm, and particle swarm optimization algorithm parameters.

3. Parameters

The most widely used validity criteria are introduced in the following section.

4. Motivations

4.1 Validity criteria

These validity criteria have been utilized for estimating the bunches. Root-Mean-Square Standard Deviation (RMSSTD), R-square, Sum of Squared Error (SSE), Internal and External legitimacy criteria applied to the previously mentioned calculations to investigate the best calculations. Bunching Algorithms utilize these approval measures to assess the outcomes. The RMSSTD is the technique to assess the change of the bunches and it gauges the group's homogeneity. According to these outcomes, to perceive homogeneous gatherings as the most minimal RMSSTD esteem implies great bunching. To gauge the divergence of bunches R-squared record is utilized. R-square estimates the level of homogeneity between the gatherings. The scope of these qualities is 0 and 1. Here, 0 methods have no distinction between the bunches and 1 method there is a huge contrast between the groups. The Sum of Squared Error is a fundamental calculation for factual methodologies and handles another estimation of information. It recognizes how those qualities are firmly related. Once figure the estimation of SSE for a dataset than just ascertain the estimations of change and standard deviation. Inner Validity is the legitimacy measure for the level of traits of free factor and others. Outer Validity is the legitimacy measure to the degree the after-effects of a summed-up study [16]. The informational collections have been taken from different assets and the subtleties of informational collections and calculations as demonstrated as follows. Sack of words informational collection have taken from UCI Machine Repository site. This informational collection is content sort, 8lakhs of occurrences, and 1 lakh of information traits. Right now, every assortment of content contains the Number of archives spoke to by D; the Number of words spoken to by W, and the Total number of words spoken to by N in the assortment.

5. Proposed work

The results of the above exploratory survey proposed to pick k-means, Leader, and ISODATA from parceling calculations and actualized on seat mark dataset with the previously mentioned legitimacy criteria for dissecting the presentation. By utilizing some developmental calculations, for example, Genetic Algorithms, Particle Swarm Optimization, and Social Group Optimization to be assessed the presentation with some legitimacy capacities. The accompanying table speaks to the subtleties of grouping strategies. Different clustering methods details with various parameters as shown in **Table 3**.

Algorithm name	Type of data handle	Time complexity	Input parameters
Leader	Numerical	O(n)	• Distance Threshold
K-means	Numerical	O(n)	• Number of Clusters
ISODATA	Numerical	O(kn)	• Minimum Number of Objects in Cluster • Possible number of Clusters • most extreme spread parameter for Splitting Maximum separation partition for Merging Maximum number of Clusters that can be combined

Table 3.
Clustering methods details.

6. Conclusion

The paper titled " Clustering Algorithms: An Exploratory Review" outlined a few dividing calculations and Evolutionary Algorithms. Apportioning Algorithms, for example, k-implies, k-medoids, Fuzzy k-means, and Expectation Maximization, etc., are considered. According to the correlation of k-implies, Fuzzy k-means, and k-medoids: The primary expert of k-implies is less expense of calculation, albeit con is empathy to Noisy information and Outliers than Fuzzy k-means and k-medoids. In Evolutionary Algorithms: GA, PSO, SGO, CSA, and TLBO are read, and for certain calculations like GA, CSA, and PSO what are the potential parameters utilized for correlations. The legitimacy criteria like RMSSTD, R-square, SSE, interior, and outside criteria have been utilized for the execution of the benchmark informational index. These legitimacy measures have been assessed for different info datasets and look at the effectiveness of the legitimacy measures.

The previously mentioned calculations actualized on seat mark informational collection with legitimacy measures to assess the presentation. In the future, by utilizing this to be evaluated execution present some new developmental calculation which can be utilized for huge and semi-organized information.

Author details

R.S.M. Lakshmi Patibandla* and Veeranjaneyulu N
Department of IT, Vignan's Foundation for Science Technology and Research, Vadlamudi, Guntur, Andhra Pradesh, India

*Address all correspondence to: patibandla.lakshmi@gmail.com

References

[1] Yujie Zheng, "Clustering Methods in Data Mining with its Applications in High Education," International Conference on Education Technology and Computer, 2012.

[2] Prabhdip Kaur, Shruti Aggrwal, "Comparative Study of Clustering Techniques," international journal for advance research in engineering and technology, April 2013.

[3] H. Men'endez and D. Camacho, "A genetic graph-based clustering algorithm," in Intelligent Data Engineering and Automated Learning -IDEAL 2012, ser. Lecture Notes in Computer Science, H. Yin, J. Costa,and G. Barreto, Eds. Springer Berlin / Heidelberg, vol. 7435,pp: 216-225, 2012.

[4] Patibandla, R.S.M.L., Veeranjaneyulu, N. (2018), "Performance Analysis of Partition and Evolutionary Clustering Methods on Various Cluster Validation Criteria", Arab J Sci Eng,Vol.43, pp.4379-4390.

[5] Y. Li, J. Chen, R. Liu, and J. Wu, "A spectral clustering-based adaptive hybrid multi-objective harmony search algorithm for community detection," in Evolutionary Computation (CEC), IEEE Congress on. IEEE2012, pp. 1-8,2012.

[6] H. Men'endez, D. F. Barrero, and D. Camacho, "A multi-objective genetic graph-based clustering algorithm with memory optimization," in 2013 IEEE Conference on Evolutionary Computation, vol. 1, pp: 3174-3181, June2013.

[7] J. Liu, W. Zhong, H. A. Abbass, and D. G. Green, "Separated and overlapping community detection in complex networks using multiobjective evolutionary algorithms," in Evolutionary Computation (CEC), 2010 IEEE Congress on. IEEE, pp: 1-7, 2010.

[8] Suresh Satapathy and Anima Naik "Social Group Optimization (SGO): a new population evolutionary optimization technique", Journal of complex intelligent systems, Springer, Vol 2, Issue 4, pp: 173-203, 2016.

[9] R S M Lakshmi Patibandla and N. Veeranjaneyulu, (2018), "Explanatory & Complex Analysis of Structured Data to Enrich Data in Analytical Appliance", International Journal for Modern Trends in Science and Technology, Vol. 04, Special Issue 01, pp. 147-151.

[10] Rao RV, "Teaching-learning-based optimization: a novel method for constrained mechanical design optimization problems," Elsevier Comput Aided Des 43, pp: 303-315,2011.

[11] R S M Lakshmi Patibandla, Santhi Sri Kurra, Ande Prasad and N.Veeranjaneyulu, (2015), "Unstructured Data: Qualitative Analysis", J. of Computation In Biosciences And Engineering, Vol. 2,No.3,pp.1-4.

[12] Wen-JyeShyr, "Introduction and Comparison of Three Evolutionary-Based Intelligent Algorithms for Optimal Design," Third International Conference on Convergence and Hybrid Information Technology, 2008.

[13] Patibandla R.S.M.L., Veeranjaneyulu N. (2018), "Survey on Clustering Algorithms for Unstructured Data". In: Bhateja V., Coello Coello C., Satapathy S., Pattnaik P. (eds) Intelligent Engineering Informatics. Advances in Intelligent Systems and Computing, vol 695. Springer, Singapore

[14] Ashish Goel, "A Study of Different Partitioning Clustering Technique," IJSRD - International Journal for Scientific Research & Development, Vol. 2, Issue 08, ISSN (online): 2321-0613, 2014.

[15] Wen-Jye Shyr, "Introduction and Comparison of Three Evolutionary-Based Intelligent Algorithms for Optimal Design," Third International Conference on Convergence and Hybrid Information Technology, 2008.

[16] R S M Lakshmi Patibandla, Veeranjaneyulu,N.(2020), "A SimRank based Ensemble Method for Resolving Challenges of Partition Clustering Methods", Journal of Scientific & Industrial Research,Vol. 79, pp. 323-327.

Chapter 3

Clustering by Similarity of Brazilian Legal Documents Using Natural Language Processing Approaches

Raphael Souza de Oliveira
and Erick Giovani Sperandio Nascimento

Abstract

The Brazilian legal system postulates the expeditious resolution of judicial proceedings. However, legal courts are working under budgetary constraints and with reduced staff. As a way to face these restrictions, artificial intelligence (AI) has been tackling many complex problems in natural language processing (NLP). This work aims to detect the degree of similarity between judicial documents that can be achieved in the inference group using unsupervised learning, by applying three NLP techniques, namely term frequency-inverse document frequency (TF-IDF), Word2Vec CBoW, and Word2Vec Skip-gram, the last two being specialized with a Brazilian language corpus. We developed a template for grouping lawsuits, which is calculated based on the cosine distance between the elements of the group to its centroid. The Ordinary Appeal was chosen as a reference file since it triggers legal proceedings to follow to the higher court and because of the existence of a relevant contingent of lawsuits awaiting judgment. After the data-processing steps, documents had their content transformed into a vector representation, using the three NLP techniques. We notice that specialized word-embedding models—like Word2Vec—present better performance, making it possible to advance in the current state of the art in the area of NLP applied to the legal sector.

Keywords: legal, natural language processing, clustering, TF-IDF, Word2Vec

1. Introduction

In recent years, the Brazilian Judiciary has been advancing toward turning all its acts digital. Following this direction, the Brazilian Labour Court implemented in 2012 the Electronic Judicial Process (acronym in Portuguese for "*Processo Judicial Eletrônico*"—PJe), and from this date, all new legal proceedings have already been born electronic. According to the Annual Analytical Report of Justice in Numbers 2020 (base year 2019) [1], produced by the National Council of Justice (acronym in

Portuguese for "*Conselho Nacional de Justiça*"—CNJ), more than 99% of the ongoing cases are already on this platform.

Knowing that human beings cannot promptly analyze a large set of data, especially when such data do not appear to correlate, a way to assist in the pattern-recognition process is through statistical, computational, and data analysis methods. From the perspective that an exponential increase in textual data exists, the analysis of patterns in legal documents has become increasingly challenging.

Currently, one of the major challenges in the legal area is to respond quickly to the growing judicial demand. The Brazilian legal system provides for ways to ensure the swift handling of judicial proceedings, such as the principle of the reasonable duration of a case, the principle of speed, the procedural economy, and due process to optimize the procedural progress [2]. Therefore, with the aid of some clustering mechanism, that is, the grouping of processes, with a good rate of similarity between the documents to be analyzed, it was possible to help in the distribution of work among the advisors of the office for which the process was drawn. In addition, it contributed to the search for case law[1] for the judgment of the cases in point, to ensure a speedy trial, upholding the principle of legal certainty. According to Gomes Canotilho [3]:

> *"The general principle of legal certainty in a broad sense (thus encompassing the idea of trust protection) can be formulated as follows: the individual has the right to be able to rely on the law that his acts or public decisions involved in his rights, positions or legal relations based on existing legal norms and valid for those legal acts left by the authorities on the basis of those rules if the legal effects laid down and prescribed in the planning are connected to the legal effects laid down and prescribed in the legal order" (2003, p. 257).*

Thus, this legal management tool created positive impacts such as the decrease of the operational costs of a legal proceeding, as a result of reducing its duration, meaning lower expenses on the allocation of the necessary resources for its judgment.

Recently, machine learning algorithms have demonstrated through research that they are powerful tools capable of solving high-complexity problems using natural language processing (NLP) [4]. In this sense, it is possible to highlight the works of [5–9], which apply the techniques of word-embedding generation, a form of vector representation of terms, and consequently of documents, taking into account their context. The use of these word embeddings is essential when analyzing a set of unstructured data presented in the form of large-volume documents in court.

Nowadays, a specialist screens the documents and distributes among the team members the legal proceedings to be judged, setting up a deviation from the main activity of this specialist, which is the production of draft decisions. This contributed to an increase in the congestion rate (an indicator that measures the percentage of cases that remain pending solution at the end of the base year) and to the decrease in the meeting of demand index (acronym in Portuguese for "*Índice de Atendimento à Demanda*"—IAD—an indicator that measures the percentage of proceedings in downtime, compared to the number of new cases). It becomes evident in the consolidated data of the Labor Justice contained in **Table 1**, with data extracted from the Annual Analytical Report of Justice in Numbers 2020 (base year 2019) [1] produced by the National Council of Justice (CNJ).

[1] A legal term meaning a set of previous judicial decisions following the same line of understanding.

	Description	2° Degree	1° Degree	Total
Workforce				
Magistrates	Legal authority	559	3077	3636
Legal workers	Public administration employee	6911	22,785	29,696
Legal load handling				
Stockpile	Number of pending cases	792,223	3,741,548	4,533,771
New cases	Number of new cases	898,104	2,632,093	3,530,197
Judged	Number of cases judged	989,324	3,036,686	4,026,010
Closed	Number of cases with final decision	941,356	3,244,652	4,185,708
Productivity indexes				
IAD	Closed cases/new cases	104.8%	123.3%	118.6%
Congestion tax	Closed cases/(new cases + stockpile)	45.7%	53.6%	52.0%
Knowledge	Fact awareness phase	—	35.1%	35.1%
Execution	Judgment enforcement phase	—	72.7%	72.7%
Indexes per magistrate				
New cases	Average number of new cases per magistrate	1607	662	821
Workflow	Average number of cases per magistrate	3583	2794	2,927
Judged cases	Average number of cases judged per magistrate	1770	1103	1216
Closed cases	Average number of cases closed per magistrate	1684	1179	1264
Indexes per legal worker				
New cases	Average number of new cases per worker	135	83	95
Judged cases	Average number of cases judged per worker	300	351	339
Closed cases	Average number of cases closed per worker	141	148	146

Table 1.
Report of indicators of Brazilian labor justice.

This work aims, therefore, to present the degree of similarity between the judicial documents that was achieved in the inferred groups through unsupervised learning *via* the application of three techniques of NLP, namely: (i) term frequency-inverse document frequency (TF-IDF); (ii) Word2Vec with CBoW (continuous bag of words) trained for general purposes for the Portuguese language in Brazil (Word2Vec CBoW pt-BR); and (iii) Word2Vec with Skip-gram trained for general purposes for the Portuguese language in Brazil (Word2Vec Skip-gram pt-BR).

This degree of congruence signals the model's performance and is set from the average similarity measure of the grouped files, based on the similarity cosine between the elements of the group to its centroid and, comparatively, by the average cosine similarity among all the documents of the group.

Aiming to delimit the scope of this research, a dataset containing information from documents of the Ordinary Appeal Interposed (acronym in Portuguese for "*Recurso Ordinário Interposto*"—ROI) type was extracted from approximately 210,000 legal proceedings. The Ordinary Appeal Interposed was used as a reference, as this is usually the type of document that induces the legal proceedings for judgment in the higher instance (2nd degree), thus instituting the Ordinary Appeal (acronym in Portuguese for "*Recurso Ordinário*"—RO). That is a free plea, an appropriate appeal against definitive and final judgments proclaimed at first instance, seeking a review of the judicial decision drawn up by a hierarchically superior body [10].

For the present work, a literature review on unsupervised machine learning algorithms applied to the legal area was performed, using NLP, and an overview of recent techniques that use artificial intelligence (AI) algorithms in word-embedding generation. Then, we applied some methods until the results were obtained, comparing and discussing them, and finally, conclusions and future challenges were presented.

2. State-of-the-art review

Machine learning algorithms have in the most recent research demonstrated a great potential to solve high-complexity problems, which follow the categories into (i) supervised machine learning algorithms; (ii) unsupervised; (iii) semi-supervised; and (iv) by reinforcement [11]. In the context of this chapter, the literature review focused on the search for the most recent research on unsupervised machine learning or clustering algorithms applied to the legal area using NLP.

The investigation revealed that there are not many works dealing with the highlighted topic, which proves its complexity. Thus, we sought to expand the research by removing the restriction to the legal area bringing light to other publications. In [12], we discussed the content recommendation system approaches based on grouping for similar articles that used TF-IDF to perform vector transformation of the document contents and, through cosine similarity, applied k-means [13] for clustering them. In [14], the authors automatically summarized texts using TF-IDF and k-means to determine the document's textual groups used to create the abstract. Then, TF-IDF is considered the primary technique for vectorizing textual content and k-means the most used algorithm for unsupervised machine learning.

Therefore, we can assume that choosing the best technique of generating word embeddings requires investigation, experimentation, and comparison of models. Several recent pieces of research have demonstrated the feasibility of using word embeddings to improve the quality of AI algorithm results for pattern detection, classification, among other uses.

In 2013, Mikolov et al. [6] proposed two new architectures to calculate vector representations of words calling them Word2Vec, which was considered, at the time, as a reference in the subject. Subsequently, techniques of word embeddings based on the use of the long short-term memory network (LSTM) [15] became widely used for speech recognition, language modeling, sentiment analysis, and text prediction, and that, unlike the recurrent neural network (RNN) they can forget, remember and update the information thus taking a step forward from the RNNs [16]. Therefore, LSTM-based libraries, such as Embeddings from Language Models (Elmo) [17], Flair [18], and context2vec [19] created a different word embedding for each occurrence of the word, related to the context, that allowed to capture the meaning of the word.

In more recent years, new techniques of word embeddings have emerged, with emphasis on (i) Bidirectional Encoder Representations from Transformers (BERT) [9], context-sensitive model with architecture based on a transformer model [20]; (ii) Sentence BERT (SBERT) [21], a "Siamese" BERT model that was proposed to improve BERT's performance when seeking to obtain the similarity of sentences; and (iii) Text-to-Text Transfer Transformer (T5) [22], a framework for treating NLP issues as a text-to-text problem, that is, input to the template as text and template output as text.

From this analysis, it was possible to advance in the current state of the art in the area of NLP applied to the legal sector, by conducting a comparative study and application of the techniques TF-IDF, Word2Vec CBoW, and Word2Vec Skip-gram to perform the grouping of labor legal processes in Brazil using the k-means algorithm and the cosine similarity.

3. Methodology

This section presents each step necessary to achieve the results and to make it possible to analyze them comparatively. To perform all the implementations of the routines necessary for this study, the Python programming language (version 3.6.9) was used and, among other libraries, (i) Numpy (version 1.19.2) was used; (ii) Pandas (version 1.1.3); (iii) Sklearn (version 0.21.3); (iv) Spacy (version 2.3.2); and (v) Nltk (version 3.5).

Every processing flow (pipeline) consists of the phases: (i) data extraction; (ii) data cleansing; (iii) generation of word-embedding templates; (iv) calculation of the vector representation of the document; (v) unsupervised learning; and (vi) calculation of the similarity measure, as detailed in the following subsections.

3.1 Data extraction

The dataset used for these studies belongs to the Regional Labour Court of the 5th Region (acronym in Portuguese for "Tribunal Regional do Trabalho da 5ª Região"—TRT5). There are approximately 210 (two hundred and ten) thousand documents of the Ordinary Appeal Interposed type, incorporated into the Electronic Judicial Process (PJe) system, originally added to the PJe in portable document format (PDF) or hypertext markup language (HTML). As the PJe has a tool for extracting and storing the contents of documents, there was no need for further processing in obtaining the text of such files.

In addition to the content of the documents, the following information was extracted: (i) the name of the parts of the proceedings to which such documents belonged; (ii) the list of labor justice issues from the Unified Procedural Table[2] (acronym in Portuguese for "*Tabela Processual Unificada*"—TPU) of the Labour Justice branch (made available by the National Council of Justice [CNJ] and consolidated by the Superior Labour Court [acronym in Portuguese for "*Tribunal Superior do Trabalho*"—TST]); and (iii) list of abbreviations (acronyms) with their full translation according to tables made available by the Supreme Court (acronym in Portuguese for "Supremo Tribunal Federal"—STF).[3]

[2] Labour Justice Unified Procedural Table. Available at: https://www.tst.jus.br/web/corregedoria/tabelas-processuais

[3] Table of abbreviations (and acronyms) made available by the Supreme Court. Available at: https://www.stf.jus.br/arquivo/cms/publicacaoLegislacaoAnotada/anexo/siglas_cf.pdf

3.2 Data cleaning

Preprocessing is a fundamental step for the application of artificial intelligence techniques and involves the following: (i) data standardization (when there is a large discrepancy between the values presented to the technique); (ii) the withdrawal of null values; and (iii) the reorganization and adequacy of the structure of the dataset. In this case, it is usually necessary for experts to conduct an exploratory analysis of the data used in advance to determine the direction of preprocessing.

For this phase, this study uses two forms of preprocessing: (i) detection of the subjects of the Unified Procedural Table (contained in the extracted documents) and (ii) cleaning the contents of the documents.

For the detection of the subjects of the TPU present in the extracted documents, regular expression matching was used as the search technique to measure the occurrences of these words in the files marking them with "tags" referring to the subject found.

For cleaning the contents of documents, usually using a regular expression, the steps were as follows:

- HTML tags: removed the html tags found in the document, such as <script>, <body>, <style> etc.;

- TPU subjects: replaced the subject text with a subject tag, for example, "*hora extra*" (overtime) changed to *hora_extra*;

- Related Persons: replaced the name of the individuals linked to the legal cases of the documents, such as the name of the author(s) and defendant(s), by the "tag" "*parteprocesso*" (part in the process);

- Judicial process number: replaced the number of the judicial process (according to the standard formatting defined nationally by the CNJ, NNNNNNN-NN. NNNN.N.NN.NNNN where N is a numeral) by the "tag" "*numeroprocesso*" (process number);

- Standardization of abbreviations: replacement of abbreviations (acronyms) by the full translation as drawn STF list as reported in Section 3.1, for example, CLT was transformed into "*Consolidação das Leis do Trabalho*" (Consolidated Labour Law);

- Addresses: replaced the addresses contained in the document with the "tag" "*enderecoprocesso*" (addresses in the process);

- Links: removed Internet links contained in the text;

- Date and Time: replacement of date and time content with "*datahora*" (datetime) tag;

- Time: replacement of the time content with the "*hora*" (hour) tag;

- Days of the week: removed the days of the week found in the document;

- Document ids: replacement of PJe document ids referenced in the document with "tag" "*sequenciadocumento*" (document sequence). These ids are typically composed of alphanumeric characters;

- Unit of measure: replaced the units of measurements and their values by the "tag" "*unidademedida*" (measurement unit);

- Numbers: replaced the numbers in full, ordinal numbers, and numerical sequences by the "tag" "*numeral*" (number);

- Judging bodies: replaced the judging bodies (e.g., "*Tribunal Regional do Traabalho*" [Regional Labour Court]) by the "*orgaojulgador*" (organjudge) tag;

- Months of the year: removed the months of the year found in the document;

- Judicial Stopwords: only when the technique employed is TF-IDF. The common words were removed in all texts of the judiciary, such as (i) "*magistrado*" (magistrate) and (ii) "*processo*" (legal proceeding), among others;

- Stopwords:

    - TF-IDF: removed all stopwords from the Portuguese language, such as "*de*" (from), "*da*" (of), "*a*" (the), "*o*" (the), "*esta*" (this) etc.;

    - Other techniques: removed only the non-adverbs of the Portuguese language, for example, the words "*não*" (no), "*mais*" (more), "*quando*" (when), "*muito*" (very), "*também*" (also), and "*depois*" (after) remain in the document;

- Line breaks: replaced line breaks by space;

- Punctuation marks:

    - TF-IDF: removed all the punctuation marks contained in the documents;

    - Other techniques: removed the punctuation marks except dot (.), comma (,), exclamation (!), and interrogation (?);

- Lemmatization:

    - TF-IDF: applied the technique to replace words with its root, for example, words such as "*tenho*" (have), "*tinha*" (had), and "*tem*" (have) had belong of the same root "*ter*" (have);

    - Other techniques: lemmatization has not been applied;

In addition to the preprocessing detailed above, when the technique used was TF-IDF, the tags inserted in the text during this phase were removed.

3.3 Generation of word-embedding templates

An essential technique in solving machine learning problems, involving NLP, is the use of vector representation of words, in which numerical values indicate some correlation of words in the text. This chapter uses word embeddings generated and shared for the Portuguese language, such as Word2Vec CBoW and Word2Vec template with Skip-gram. These templates were created based on more than 1 billion and 300,000 tokens, with results published in the article "Portuguese Word Embeddings: Evaluating on Word Analogies and Natural Language Tasks" presented at the Symposium in Information and Human Language Technology - STIL 2017 [23].

3.4 Calculation of the vector representation of the document

Different from the TF-IDF technique, which has the vector representation of the document based on the statistical measurement of each term of the document in relation to all known corpus, and whose vector dimension is equal to the size of the vocabulary of the corpus, the other techniques (i) Word2Vec CBoW ptBR and (ii) Word2Vec Skip-gram pt-BR need to go through a change to calculate the vector representation of the document (document embeddings). This happens because for these techniques what you can get is the vector representation of the word (word embeddings).

Thus, to calculate the vector representation for the documents some alternatives are suggested, such as (i) average of the word embeddings of the words of the document; (ii) sum of the word embeddings of the words in the document by pondering them with the TF-IDF and then dividing by the sum of the TF-IDF of the words of the document; and (iii) weighted average with the TF-IDF of the word embeddings of the words of the document, the latter being the technique chosen for presenting the best result.

3.5 Unsupervised learning

The use of unsupervised learning techniques is relevant when the intention is to detect patterns among court documents. The k-means algorithm, whose basic concepts were proposed by MacQueen [13], is the technique adopted in this study. In general, this technique seeks to recognize patterns from the random choice of K initial focal points (centroid), where K is the number of groups that one wishes to obtain and, iteratively, position the elements whose Euclidean distance is the minimum possible concerning the centroid of the group.

Since one does not have an ideal K to offer the algorithm, an approach usually used to support such a decision is to calculate the inertia, based on how well the dataset was grouped through k-means.

The inertia calculation is based on the sum of the square of the Euclidean distance from each point to its centroid and seeks to obtain the lowest K with the lowest inertia. However, the higher the K value reaches, the tendency is that inertia will be lower, and then, the elbow method was used to find the point where the reduction in inertia begins to decrease.

Hence, 31 values for K were used within the range from 30 to 61, considering an interval for each unit, selecting the K that generated the best grouping. In addition, the strategy of creating submodels, limited to two, was used for the documents of the groups whose average similarity rate did not reach a value greater than 0.5.

3.6 Similarity measure calculation

The similarity measure is an important tool for the measurement of the quality of inferred groups. In this study, the cosine similarity measure is adopted, which is a measure that calculates the cosine of the angle between two vectors projected in the multidimensional plane, the result of which is between 0 and 1, in which 1 represents that the two vectors are totally similar, and 0 represents that they are totally different. Given two vectors, X and Y, the cosine similarity is presented using a scalar product according to Eq. (1).

$$\text{similarity} = \cos(\theta) = \frac{X \cdot Y}{|X| \cdot |Y|} \quad (1)$$

Consequently, to decide whether, after the clustering of the chief model, it was necessary to generate up to two more submodels, using the average cosine similarity among all elements of the group. Although the computational cost of calculating the similarity between all files in the group is relevant, we sought to reduce the distance between documents that were part of the same group, although they were located near the centroid. To assess the final efficiency of the technique, another form of calculation was adopted, computing for each group the average cosine similarity between the group elements and its centroid. Thus, as a measure of global similarity of each approach, we calculated the average of the average of the groups, so that the one that reached a value closer to 1 (one) was considered the best technique.

4. Results and discussions

This research shows, as per the methodology presented in the previous sections, how machine learning algorithms associated with NLP techniques are important allies in optimizing the operational costs of the judicial process. It is evidenced from the result, for example, of document screenings and procedural distribution, which allows an expert to devote oneself to their chief activity optimizing working time.

While using the k-means unsupervised learning algorithm, it was necessary to choose the best K for each NLP technique studied. In this scenario, the elbow method was applied based on the calculated inertia of each of the 31 K tested, as shown in **Figure 1**, thus achieving a better result for each technique.

From the attainment of the best K, the k-means model was trained and, from the grouping performed by this technique, we could reach the average similarity between the documents of each group. Those groups that did not make the cutting line of at least 0.5 of average had the group files submitted for creating up to two submodels. As expected, only for TF-IDF technique groupings is there a need to generate submodels to improve performance.

Table 2 shows the average similarity of the groups obtained using the TF-IDF technique, as well as the result of the Word2Vec CBoW pt-BR technique. It achieved a little better measure of similarity than the Word2Vec Skip-gram pt-BR technique; however, the latter achieved its result with a smaller number of groups, which places it, in general, as the best technique.

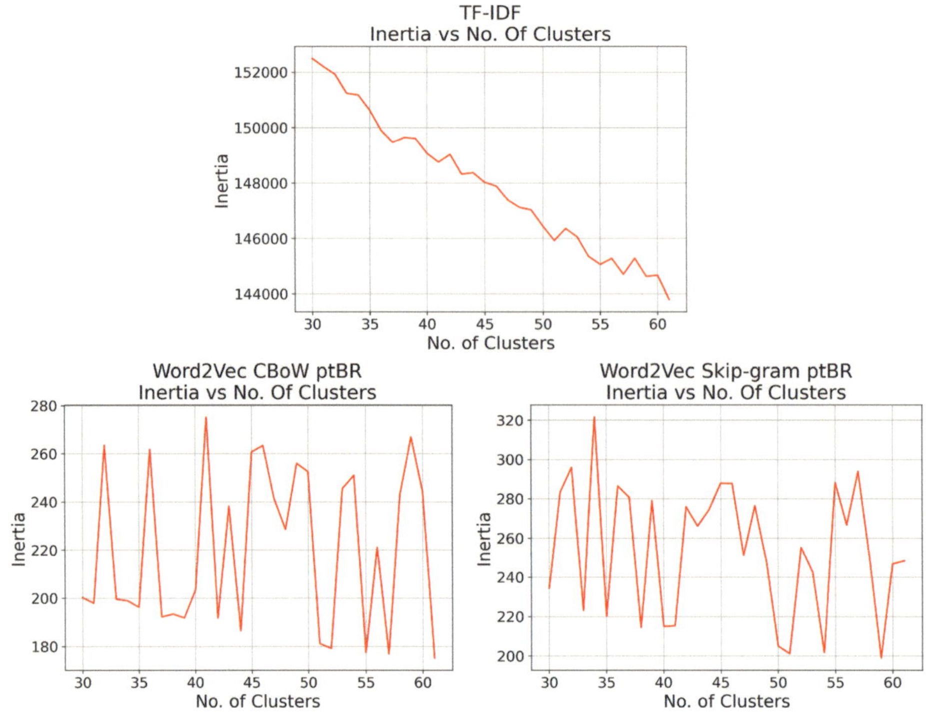

Figure 1.
Inertia charts constructed by using the elbow method for determining the best number of clusters for each approach.

	Model		Submodel 1		Submodel 2		Final	
Type	**Groups**	**Mean**	**Groups**	**Mean**	**Groups**	**Mean**	**Groups**	**Mean**
TF-IDF	37	0.3696	43	0.4001	48	0.4002	48	0.4002
Word2Vec CBoW ptBR	59	0.9060	—		—		59	0.9060
Word2Vec Skip-gram ptBR	34	0.9044	—		—		**34**	**0.9044**

Table 2.
Mean cosine similarity between all elements of the group. The best results are highlighted in bold.

After the groups were formed, the statistical data resulting from each approach were calculated, as shown in **Table 3** and in the comparative graph of distributions between the techniques (**Figure 2**). The cosine similarity of the group elements to its centroid was used as a metric, showing the proximity of the results between the techniques with Word2Vec and highlighting the technique Word2Vec Skip-gram ptBR for the smaller amount of generated groups.

When comparing the values presented in **Tables 2** and **3**, it is noteworthy that the results presented in **Table 2** are worse in all cases. It is inferable from this observation

Type	Groups	Mean	Std.	Min.	25%	50%	75%	Max.
TF-IDF	49	0.6241	0.1718	0.2466	0.5021	0.5864	0.1639	0.9644
Word2Vec CBoW ptBR	59	0.9475	0.0632	0.7640	0.9352	0.9790	0.991	0.9999
Word2Vec Skip-gram ptBR	**34**	**0.9481**	**0.0609**	**0.7960**	**0.9248**	**0.9763**	**0.9924**	**0.9995**

Table 3.
Statistics of the cosine similarity of the group elements to the centroids. The best results are highlighted in bold.

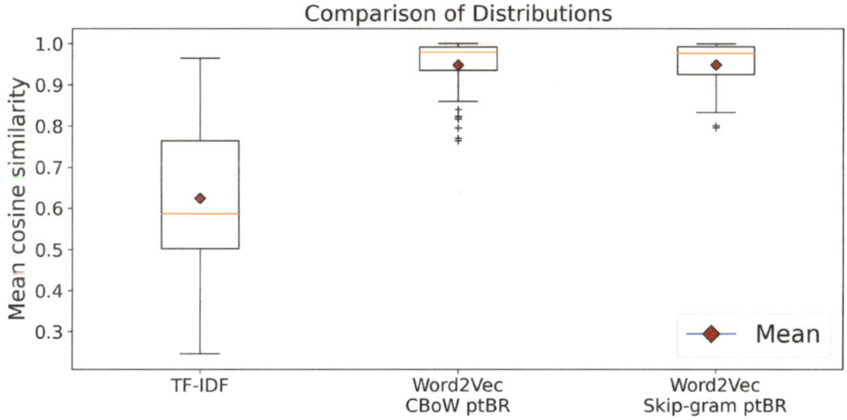

Figure 2.
Boxplots showing the distributions of the clusters calculated by each technique. The more cohesive the boxes and the less number of outliers, the better.

that the similarity measure calculations shown in **Table 2** can reduce the similarity rates since there may be elements in the group positioned on completely opposite sides. From **Figure 2**, it is also possible to verify that the groupings generated by the Word2Vec technique were more cohesive than those generated by the TF-IDF technique, especially the Word2Vec Skip-gram technique, which created fewer groupings in the range of outliers than Word2Vec CBoW, demonstrating its superiority by allowing fewer groups but maintaining consistent quality and cohesion.

Given the aforesaid, among all the techniques evaluated, the Word2Vec Skip-gram pt-BR technique presented itself as the best option for word embeddings for clustering legal documents of the Ordinary Appeal Interposed type. Although the Word2Vec CBoW pt-BR technique achieves slightly better rates, it stands out from the previous one for reaching a much smaller number of groups.

The result achieved by each approach can be visualized by projecting in two dimensions of the groups formed from the three techniques: (i) TF-IDF; (ii) Word2Vec CBoW pt-BR; and (iii) Word2Vec Skip-gram pt-BR, respectively, presented in **Figures 3–5**. It is evident in the figures that the groups formed from Word2Vec are much better defined, especially skip-gram, which confirms the findings previously explained in this work.

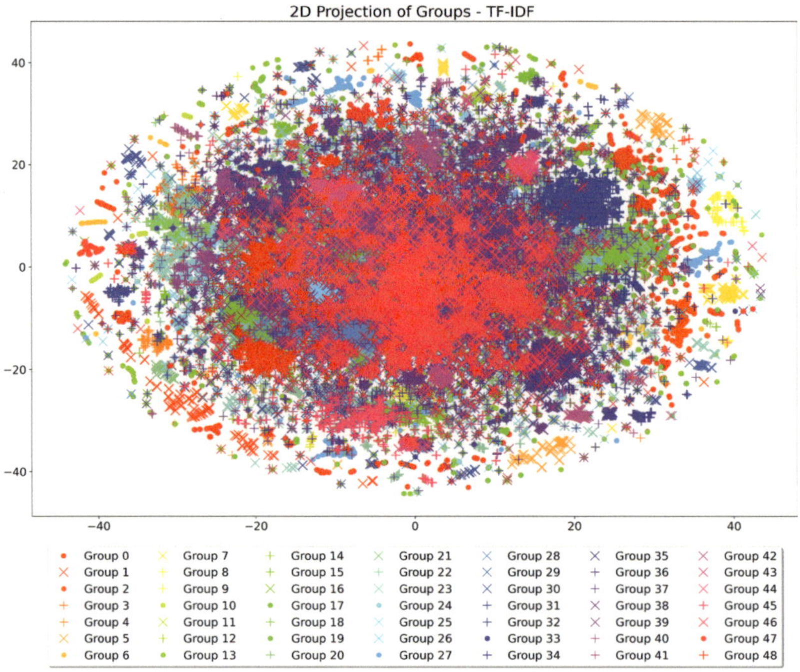

Figure 3.
2D projection of the entire test dataset, showing for each document its corresponding group formed by TF-IDF.

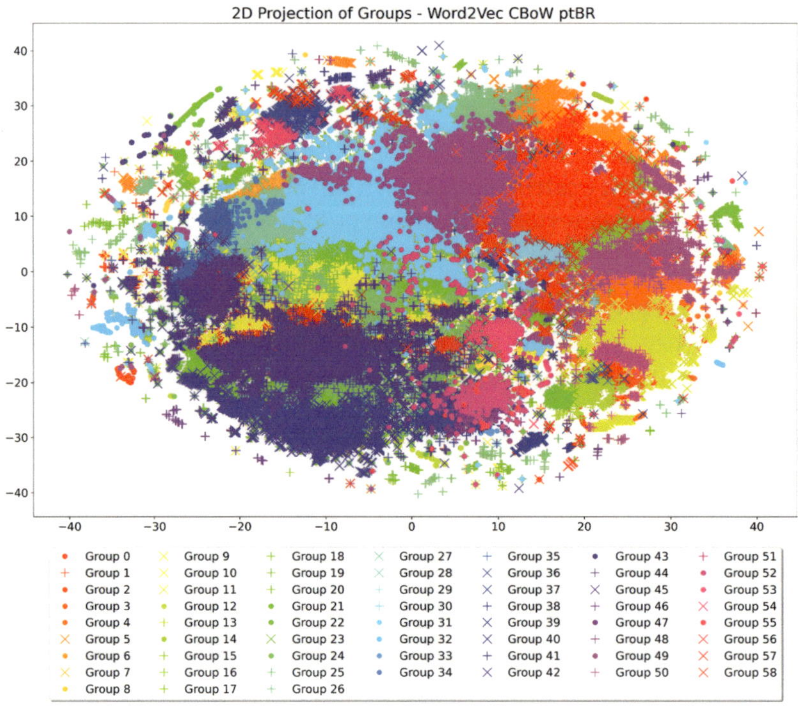

Figure 4.
2D projection of the entire test dataset, showing for each document its corresponding group formed by Word2Vec CBoW ptBR.

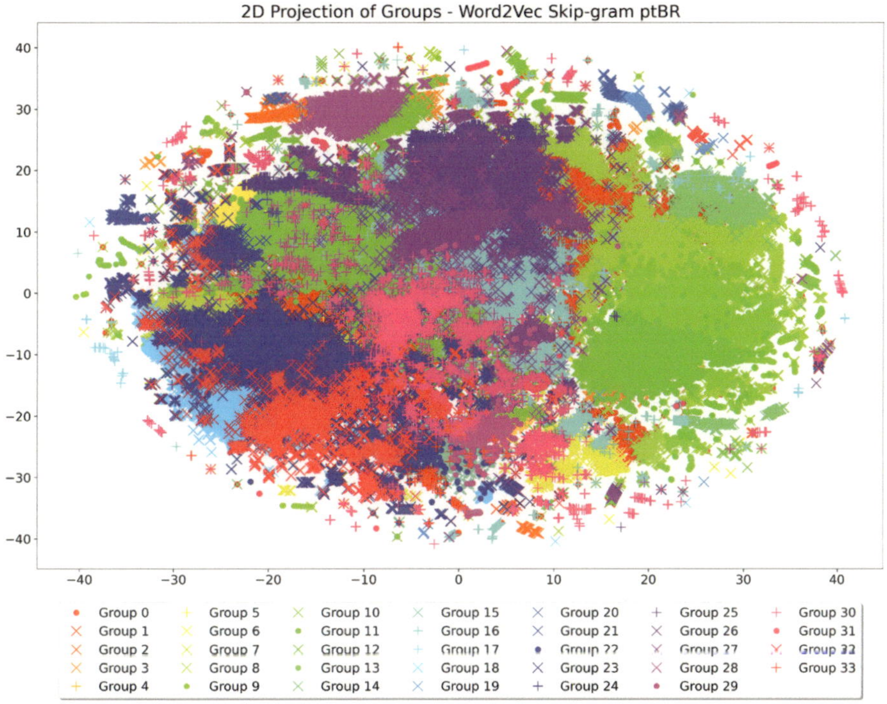

Figure 5.
2D projection of the entire test dataset, showing for each document its corresponding group formed by Word2Vec skip-gram ptBR.

5. Conclusion and future work

The use of AI as a standard detection tool based on documents from the judiciary has generally proved to be a viable and helpful solution in the scientific, technological, and practice of legal work. In this chapter, it was possible to present the results considered very promising due to the improvement in the average similarity rate. Thus, we demonstrate the possibility of using word-embedding generation techniques applied on clustering of Ordinary Appeal Interposed using AI algorithms.

Of all the techniques evaluated, the Word2Vec Skip-gram pt-BR technique presented itself as the best option for word embeddings for clustering legal documents of the Ordinary Appeal Interposed type.

We believe that specialized word embeddings have great potential in improving the results. Therefore, comes the suggestion for future study of Word2Vec specialized for the judiciary, in addition to evaluating whether the new embeddings generated provide an opportunity to improve the overall performance of clustering. In addition, using transformer-based techniques, such as BERT, can achieve promising results, using both the Portuguese language word-embedding model and training a specialized BERT model for the judiciary.

Moreover, new possibilities arise for using the techniques discussed in this chapter, such as the draft generation of decisions and classification of documents and processes.

Acknowledgements

The authors thank the Regional Labour Court of the 5th Region for making datasets available to the scientific community and contributing to research and technological development. The authors also thank the Artificial Intelligence Reference Centre and the Supercomputing Centre for Industrial Innovation, both from SENAI CIMATEC.

Author details

Raphael Souza de Oliveira[1] and Erick Giovani Sperandio Nascimento[2*]

1 TRT5—Regional Labor Court of the 5th Region, Salvador, BA, Brazil

2 SENAI CIMATEC—Manufacturing and Technology Integrated Campus, Salvador, BA, Brazil

*Address all correspondence to: ericksperandio@gmail.com

References

[1] CNJ—Conselho Nacional de Justiça. Relatório Analítico Anual da Justiça em Números 2020. 2020. Available from: https://www.cnj.jus.br/pesquisas-judiciarias/justica-em-numeros/ [Accessed: June 07, 2021]

[2] da Costa Salum G. A duração dos processos no judiciário: aplicação dos princípios inerentes e sua eficácia no processo judicial [Internet], Âmbito Jurídico, Rio Grande. Vol. XIX(145). 2016. Avaliable from: https://ambitojuridico.com.br/cadernos/direito-processual-civil/a-duracao-dos-processos-no-judiciario-aplicacao-dos-principios-inerentes-e-sua-eficacia-no-processo-judicial/ [Accessed: September 01, 2021]

[3] Canotilho JJG. Direito constitucional e teoria da constituição. 7th ed. Coimbra: Almedina; 2003

[4] Khan W, Daud A, Nasir J, Amjad T. A survey on machine learning models for Natural Language Processing (NLP). Computer Science and Engineering. 2016;**43**:95-113

[5] Wang Y, Cui L, Zhang Y. Using Dynamic Embeddings to Improve Static Embeddings. In: arXiv Preprint. arXiv:1911.02929v1. 2019

[6] Mikolov, T, Chen, K, Corrado, G, Dean, J. Efficient Estimation of Word Representations in Vector Space. In: ICLR: Proceeding of the International Conference on Learning Representations Workshop Track, Arizona, USA. 2013.

[7] Pennington J, Socher R, Manning C. Glove: Global vectors for word representation. In: Proceedings of the 2014 Conference on Empirical Methods in Natural Language Processing (EMNLP). Doha, Qatar: Association for Computational Linguistics; 2014. pp. 1532-1543

[8] Bojanowski P, Grave E, Joulin A, Mikolov T. Enriching word vectors with subword information. Transactions of the Association for Computational Linguistics. 2017;**5**:135-146

[9] Devlin J, Chang M-W, Lee K, Toutanova K. BERT: Pre-training of Deep Bidirectional Transformers for Language Understanding. In: Proceedings of the 2019 Conference of the North American Chapter of the Association for Computational Linguistics: Human Language Technologies. Minneapolis, Minnesota: Association for Computational Linguistics. 2019; 1:4171-4186. DOI: 10.18653/v1/N19-1423

[10] Oliveira FJV. Os recursos na Justiça do Trabalho [Internet]. Available from: http://www.conteudojuridico.com.br/consulta/Artigos/24853/os-recursos-na-justica-do-trabalho [Accessed: June 10, 2021]

[11] Sil R, Roy A, Bhushan B, Mazumdar AK. Artificial Intelligence and Machine Learning based Legal Application: The State-of-the-Art and Future Research Trends. In: 2019 International Conference on Computing, Communication, and Intelligent Systems (ICCCIS); 18-19 October 2019; Greater Noida, India: IEEE; 2019. p. 57-62. DOI: 10.1109/ICCCIS48478.2019.8974479

[12] Renuka S, Raj Kiran GSS, Rohit P. An unsupervised content-based article recommendation system using natural language processing. In: Jeena Jacob I, Kolandapalayam Shanmugam S, Piramuthu S, Falkowski-Gilski P, editors. Data Intelligence and Cognitive

Informatics (Algorithms for Intelligent Systems). Singapore: Springer; 2021. pp. 165-180. DOI: 10.1007/978-981-15-8530-2_13

[13] MacQueen J. Some methods for classification and analysis of multivariate observations. In: Proceedings of the Fifth Berkeley Symposium on Mathematical Statistics and Probability; Berkeley, CA: University of California Press; Vol. 1. 1967. pp. 281-297.

[14] D'Silva J, Sharma U. Unsupervised automatic text summarization of Konkani texts using K-means with Elbow method. International Journal of Engineering Research and Technology. 2020;**13**:2380. DOI: 10.37624/IJERT/13.9.2020.2380-2384

[15] Hochreiter S, Schmidhuber J. Long short-term memory. Neural Computation. 1997;**9**:1735-1780. DOI: 10.1162/neco.1997.9.8.1735

[16] Sherstinsky A. Fundamentals of recurrent neural network (RNN) and long short-term memory (LSTM) network. Physica D: Nonlinear Phenomena. 2020;**404**:132306

[17] Peters ME, Neumann M, Iyyer M, Gardner M, Clark C, Lee K, et al. Deep contextualized word representations. In: Proceedings of the 2018 Conference of the North American Chapter of the Association for Computational Linguistics: Human Language Technologies. Vol. 1 (Long Papers). New Orleans, Louisiana: Association for Computational Linguistics; 2018. pp. 2227-2237. DOI: 10.18653/v1/N18-1202

[18] Akbik A, Blythe D, Vollgraf R. Contextual string embeddings for sequence labeling. In: Proceedings of the 27th International Conference on Computational Linguistics. Santa Fe, New Mexico, USA: Association for Computational Linguistics; 2018. pp. 1638-1649

[19] Melamud O, Goldberger J, Dagan I. context2vec: Learning Generic Context Embedding with Bidirectional LSTM. In: Proceedings of The 20th SIGNLL Conference on Computational Natural Language Learning; Berlin, Germany: Association for Computational Linguistics; 2016;. p. 51-61. DOI: 10.18653/v1/K16-1006

[20] Vaswani A, Shazeer N, Parmar N, Uszkoreit J, Jones L, Gomez AN, et al. Attention is all you need. In: Proceedings of the 31st International Conference on Neural Information Processing Systems. Red Hook, NY, USA: Curran Associates Inc.; 2017. p. 6000-10. (NIPS'17).

[21] Reimers N, Gurevych I. Sentence-BERT: Sentence Embeddings using Siamese BERT-Networks. In: Proceedings of the 2019 Conference on Empirical Methods in Natural Language Processing and the 9th International Joint Conference on Natural Language Processing (EMNLP-IJCNLP). Hong Kong, China: Association for Computational Linguistics; 2019. p. 3982-92. DOI: 10.18653/v1/D19-1410

[22] Roberts A, Raffel C, Lee K, Matena M, Shazeer N, Liu PJ, et al. Exploring the Limits of Transfer Learning with a Unified Text-to-Text Transformer. In: arXiv Preprint. arXiv:1910.10683. 2019

[23] Hartmann NS, Fonseca ER, Shulby CD, Treviso MV, Rodrigues JS, Aluísio SM. Portuguese Word Embeddings: Evaluating on Word Analogies and Natural Language Tasks. In: Proceedings of the 11th Brazilian Symposium on Information and Human Language Technology (STIL). Uberlândia, Minas Gerais, Brazil: Brazilian Computing Society - SBC; 2017. p. 122-31.

Chapter 4

Assessing Heterogeneity of Two-Part Model via Bayesian Model-Based Clustering with Its Application to Cocaine Use Data

Ye-Mao Xia, Qi-Hang Zhu and Jian-Wei Gou

Abstract

The purpose of this chapter is to provide an introduction to the model-based clustering within the Bayesian framework and apply it to asses the heterogeneity of fractional data via finite mixture two-part regression model. The problems related to the number of clusters and the configuration of observations are addressed via Markov Chains Monte Carlo (MCMC) sampling method. Gibbs sampler is implemented to draw observations from the related full conditionals. As a concrete example, the cocaine use data are analyzed to illustrate the merits of the proposed methodology.

Keywords: model-based clustering, finite mixture model, two-part model, Markov Chain Monte Carlo sampling, cocaine use data

1. Introduction

A recurring theme in the statistical analysis is to separate the unstructured data into groups to detect the similarity or discrepancy within or between groups. This is especially true in the fields, e.g., discriminant analysis [1–3], pattern recognition [4, 5], gene expression [6–8], machine learning [9], and artificial intelligence [10]. In the literature, the clustering problem is often formulated within the *cluster analysis* framework, which is generally categorized into two classes: the non-probabilistic framework and the probabilistic framework. The non-probabilistic clustering method, including the K-means method [9, 11, 12] and the hierarchical/agglomerative clustering algorithms [13–15], is based on the *distance* between any two observations or groups. It clusters data by merging or removing observations according to the "closeness" specified by the distance. This method is more general since it does not impose any distributional assumptions on data, hence having greater flexibility in the real applications. Instead, the non-probabilistic clustering algorithm, also termed the *model-based clustering*, groups data by positing a probability model on data and then clustering data via configuration function related to the model. Compared with the non-probabilistic framework, the model-based methods enable us to assess the statistical properties of the solutions, e.g., how many clusters are there, how well the configuration function works, and how robust the method is against the model deviation and so on. There is rich literature on this issue. Among them, finite mixture model (FMM, [16–18]) perhaps is the most popular choice and has often been proposed and studied in the context of clustering (see a short review in Fraley and Raftery [2]). FMM assumes that each cluster is identified with a probability distribution indexed by the cluster-specific parameter(s), and each observation is related to clusters via configuration or membership function. The statistical task is the inference about the number of clusters, the estimation of the unknown parameters, and the allocation of observations.

In this chapter, we pursue a Bayesian model-based method to address the heterogeneity of fraction data. Fractional data are very common in the social and economical surveys. A distinguished feature of fractional responses is that its measurements are responded on a scale in the unity interval [0,1] but suffer from excessive zeros and unities on the boundaries. In understanding such type of data, the commonly used method is to separate the whole data into three parts: two corresponding to the zeros and unities respectively, and one corresponding to the continuously positive values. Two separative logistic models are suggested to model two discrete value parts respectively while single normal linear regression model is formulated for the continuous value part. This method, though more appealing, ignores the instinct association across different parts and readily leads to inconsistence of the occurrence probabilities on each part. Instead, we propose a three-category multinomial model for the occurrence variable, in which the usual separated models can be considered as the marginal models of our proposal. Such modeling always ensures the probabilities on each part to be proper, thus avoiding parameter constraints, see for example, [19]. To assess the heterogeneity underlying data, we formulate the problem into a finite mixture analysis of which each component is specified by two-part regression model. In view of the model complexity, we implement Markov Chains Monte Carlo sampling method to implement posterior analysis. Block Gibbs sampler is implemented to draw observations from the target distributions. The posterior inference including parameters estimates, model selection, and the configuration determination of observations are obtained based on the simulated observations.

The chapter is organized as follows. Section 2 introduces a general model-based clustering method to address the heterogeneity of regression model within the Bayesian framework. In Section 3, we apply the proposed method to the fractional data. Section 4 presents a cocaine use study. And Section 5 concludes the chapter.

2. Method description

2.1 General framework

Suppose that for $i = 1, 2, \cdots, n$, y_i is an observed response, each associated with an m dimensional fixed covariates $\mathbf{x}_i = (x_{i1}, \cdots, x_{im})$. In the context of regression analysis, the interest mainly focuses on exploring the pattern of the influence of $\mathbf{x}_i$ on y_i and predicting the mean of a future response y in terms of a new $\mathbf{x}$. This is usually achieved by formulating $\{\mathbf{x}_i, y_i\}$ as $\mathbb{E}(y_i|\mathbf{x}_i) = m(\mathbf{x}_i)$ for some mean function $m(\cdot)$. In the parametric fitting framework, the function $m(\mathbf{x})$ is assumed to be related to $\mathbf{x}$ via linking function as the form of

$$m(\mathbf{x}) = h(\mathbf{x}^T\boldsymbol{\beta}) \tag{1}$$

which induces the so-called generalized linear model [20] for $\{\mathbf{x}_i, y_i\}$, where $\boldsymbol{\beta}$ is the regression coefficients used to quantify the uncertainty about m, and $h(\cdot)$ is the known linking function used to link the mean and the predictors.

More often, the single relationship such as Eq. (1) may not be sufficient when the patterns among the subjects take on the heterogeneity such as clustering. The heterogeneous data occur when the observations are generated from the different populations of which the number of populations and the membership of each observation to the population are unknown. The main objective is to separate data into different clusters to detect the possible similarity within clusters or the discrepancy between clusters. This is generally accomplished by defining a cluster's membership/configuration function $\mathcal{K} : \{(\mathbf{x}_1, y_1), \cdots, (\mathbf{x}_n, y_n)\} \mapsto \{1, \cdots, K\}$ such that $K_i = \mathcal{K}((\mathbf{x}_i, y_i)) = k$ if $(\mathbf{x}_i, y_i)$ belongs to the cluster k, where K is assumed to be less than n. The discrepancy between any two clusters is characterized by the cluster-specific parameters such as intercepters, regression coefficients, and/or disperse parameters.

The model-based clustering assumes that given the clusters membership K_i, $(\mathbf{x}_i, y_i)$ within the cluster k has the following sampling density

$$(y_i | K_i = k, \mathbf{x}_i) \overset{ind.}{\sim} f_k(y_i | \mathbf{x}_i^T \boldsymbol{\beta}_k, \ \tau_k) \tag{2}$$

while K_i is specified by

$$\mathbb{P}(K_i = k) = \pi_k \tag{3}$$

where f_k, maybe independent of k, is the probability density function, $\boldsymbol{\beta}_k$ and $\boldsymbol{\tau}_k$ are the cluster-specific regression coefficients and the disperse parameters, respectively, and π_k is the mixing proportion identifying the proportion of the component k over the entire population. It is assumed that $\pi_k \geq 0$ and $\sum_{k=1}^{K} \pi_k = 1.0$.

Two important issues arise when formulating data clustering problem as Eqs. (2) and (3). One is related to the number of clusters, and the other is pertained to the determination of configurations. Within the Bayesian framework, several methods have been proposed for the first issue. One can, for example, follow [21] and treat K to be random and assign a prior to it. The reversible jump MCMC method (RJMCMC, [21, 22]) can be implemented to conduct the joint analysis of K with other random quantities. Another method is along the lines with the hypothesis test procedure and routinely to estimate K via model comparison/selection procedure. This perhaps is the most popular choice in the model-based clustering context, in which various measures such as the Akaike information criterion (AIC) [23], the corrected AIC (AICc) [24, 25], the Bayesian information criterion (BIC) [26], the integrated completed likelihood (ICL) [27], and Bayes factor (BF, [28, 29]) can be adopted to select a suitable model. It is worth pointing out that the deviance information criterion (DIC) [30] may not be appropriate for the mixture model comparison. The well-known software WinBUGS® [31] for Bayesian analysis does not provide DIC results for mixture analysis. In addition, many authors suggested modeling heterogeneous data into the mixture of Dirichlet process (MDP, [32, 33]). However, as discussed in Ishwaran and James [34], DP fitting often overestimates the number of clusters and readily leads to model over fitting.

For the second issue, the complexity of problem depends on the methods adopted in the analysis. In the frequency framework, for example, the configuration of observation i is often achieved by maximizing $\mathbb{P}(K_i = k | \mathbf{Y}, \hat{\boldsymbol{\pi}}, \hat{\boldsymbol{\Xi}})$ over $k = 1, \cdots, K$, where $\hat{\boldsymbol{\pi}}$ and $\hat{\boldsymbol{\Xi}}$

are the maximum likelihood estimates (MLE) obtained via, e.g., the expectation-maximization algorithm (EM, [35]). In the next section, we will present a Bayesian procedure for determining $\mathcal{K}$. Compared with the frequency approach, the nice feature of the Bayesian approach is its flexibility to utilize prior information for achieving better results. Also, the sampling-based Bayesian methods depend less on the asymptotic theory and hence have the potential to produce reliable results even with small sample size.

Let **Y** be the set of all observed responses and **X** be the set of fixed covariates; Write Ξ as the collection of $\boldsymbol{\beta}_k$ and $\boldsymbol{\tau}_k$. Integrating over K_i produces a K-component mixture model for y_i, which is given by

$$p(y_i|\boldsymbol{\pi}, \Xi, \mathbf{x}_i) = \sum_{k=1}^{K} \pi_k f_k(y_i|\mathbf{x}_i^T \boldsymbol{\beta}_k, \boldsymbol{\tau}_k). \tag{4}$$

The log-likelihood of the observed data conditional on K is given by

$$\mathcal{L}(\pi, \Xi|\mathbf{Y}, \mathbf{X}) = \sum_{k=1}^{n} \log \left(\sum_{k=1}^{K} \pi_k f_k(y_i|\mathbf{x}_i^T \boldsymbol{\beta}_k, \boldsymbol{\tau}_k) \right). \tag{5}$$

As an illustration, **Figure 1** presents a three-component normal linear mixture regression model with one covariate. It can be seen clearly that the density function illustrates strong heterogeneity. The regression line is obviously different from those of components, which indicates that single model is unappreciate in fitting such data. In what follows, we suppress **X** for notational simplicity.

2.2 Bayesian model-based clustering via MCMC

Bayesian analysis for analyzing Eqs. (2) and (3) especially $\mathcal{K}$ requires the specification of a prior distribution $p(\boldsymbol{\pi}, \Xi)$ for the parameters of the mixture model. By model convention, it is naturally to assume that $\boldsymbol{\pi}$ and Ξ are independent, and the components among Ξ are also independent. In particular,

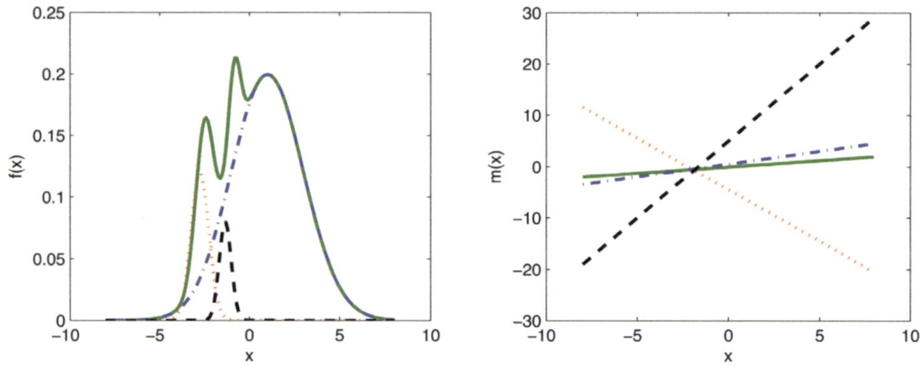

Figure 1.
Plot of the three-component normal mixture model $0.3N(-4-2x, 1) + 0.5N(0.5+0.5x, 1) + 0.2N(4.5+3x, 1)$. *Left panel: Plot of the density functions of the mixture as well as their three weighted components ; right panel: plots of regression lines. Mixture model: solid line "–" component one: dotted lines "..." component two: dashed lines "– –" and component three: dotted-dashed lines "–·"*

$$\boldsymbol{\beta}_k \overset{iid.}{\sim} N_m(\boldsymbol{\beta}_0, \boldsymbol{\Sigma}_0), \quad \boldsymbol{\tau}_k^{-1} \overset{iid.}{\sim} W(\rho_0, \boldsymbol{R}_0) \tag{6}$$

in which $W(\rho_0, \boldsymbol{R}_0^{-1})$ is the Wishart distribution with the degrees of freedom ρ_0 and the scale matrix $\boldsymbol{R}_0$, and reduces to the scaled Chi-square distribution when $\boldsymbol{\tau}_k$ is a univariate; $\boldsymbol{\beta}_0$, $\boldsymbol{\Sigma}_0$, ρ_0 and $\boldsymbol{R}_0$ are the hyper-parameters, which are treated to fixed and known. In the real applications, if no extra information can be available, the values of these hyper-parameters are often taken to ensure $\boldsymbol{\beta}_k$ and $\boldsymbol{\tau}_k$ to be dispersed enough. For example, one can set $\boldsymbol{\Sigma}_0 = \lambda_0 \mathbf{I}$ with large λ_0 (Throughout, we use $\mathbf{I}$ to signify an identify matrix). In this case, the values of $\boldsymbol{\beta}_0$ are not really important and can be set to any values, e.g., zeros. Note that for the mixture models, Diebolt and Robert [36] (see also, for example, [37]) showed that using fully non-informative prior distributions may lead to improper posterior distributions and hence is strictly prohibitive.

We assign a symmetric Dirichlet distribution to $\boldsymbol{\pi}$ as follows

$$\boldsymbol{\pi}|\alpha \sim D_K(\alpha, \cdots, \alpha) \tag{7}$$

in which $\alpha(>0)$ is the hyper-parameter, which is treated to fixed and unknown. In the applications, we can take sensitive analysis by setting smaller and larger values for α. See section 4 for more details.

Let $\mathbf{K} = \{K_1, \cdots, K_n\}$ be the collection of all configurations. A Bayesian procedure for model based clustering mainly focuses on exploring the behavior of the posterior of $\mathbf{K}$ given data, which is given by

$$p(\mathbf{K}|\mathbf{Y}) \propto p(\mathbf{Y}|\mathbf{K})p(\mathbf{K}) \tag{8}$$

where $p(\mathbf{Y}|\mathbf{K})$ is the marginal distribution of $p(\mathbf{Y}, \pi, \Xi|\mathbf{K})$ with $\boldsymbol{\pi}$ and Ξ being integrated out. Generally, no closed form can be available for this target distribution. Markov Chain Monte Carlo [38, 39] sampling method can be used to conduct posterior analysis. In particular, one can follow the routine in Tanner and Wong [40] and treat the latent quantities $\{\boldsymbol{\pi}, \mathbf{K}, \Xi\}$ as the missing data and augment them with the observed data. Posterior analysis is carried out based on the joint distribution $p(\boldsymbol{\pi}, \mathbf{K}, \Xi|\mathbf{Y})$. In this case, block Gibbs sampler [41, 42] can be implemented to draw observations from such target distribution. The Gibbs sampler is iteratively implemented by drawing: (i) Ξ from $p(\Xi|\boldsymbol{\pi}, \mathbf{K}, \mathbf{Y})$; (ii) $\boldsymbol{\pi}$ from $p(\boldsymbol{\pi}|\mathbf{K}, \Xi, \mathbf{Y})$ and $\mathbf{K}$ from $p(\mathbf{K}|\boldsymbol{\pi}, \Xi, \mathbf{Y})$ till convergence. The convergence can be monitored by the "estimated potential scale reduction" (EPSR) values [43] or by plotting the traces of estimates against iterations under different starting values. Note that except for (i), all full conditionals involved in the Gibbs sampler are standard. However, drawing Ξ in (i) depends on the specific form of the density function f_k and sometimes requires implementing Metropolis-Hastings algorithm (MH, [44, 45]) or rejection sampling [46].

2.3 Label switching

Formulating the model-based clustering problem into mixture model Eq. (2) faces the model identification. A statistical model is said to be identified if the observed likelihood is uniquely determined by unknown parameters. A less identified model may be problematic and will distort the estimates of unknown parameters. It is easily showed that the observed likelihood of data is only determined up to the permutation of the component labels. As a matter of fact, suppose that there are the pair $\{\boldsymbol{\pi}^{(1)}, \Xi^{(1)}\}$ and $\{\boldsymbol{\pi}^{(2)}, \Xi^{(2)}\}$ such that

$$p\left(y|\boldsymbol{\pi}^{(1)}, \ \boldsymbol{\Xi}^{(1)}\right) = p\left(y|\boldsymbol{\pi}^{(2)}, \boldsymbol{\Xi}^{(2)}\right) \tag{9}$$

then there exists a permutation $\nu : \{1, 2, \cdots, K\} \mapsto \{1, 2, \cdots, K\}$ such that $\pi_k^{(1)} = \pi_{\nu(k)}^{(2)}$, $\boldsymbol{\beta}_k^{(1)} = \boldsymbol{\beta}_{\nu(k)}^{(2)}$ and $\boldsymbol{\tau}_k^{(1)} = \boldsymbol{\tau}_{\nu(k)}^{(2)}$. In this setting, we can not distinguish $\mathcal{K}$ and $\nu \circ \mathcal{K}$ in terms of data ("$\circ$" denotes the operator of function composition). With this in mind, any exchangeable priors on $\boldsymbol{\pi}$ and $\boldsymbol{\Xi}$ like Eqs. (6) and (7) produces symmetric and multi-modal posterior distributions with up to $K!$ copies of each "genuine" mode, which induces the so-called label switching problem on Bayesian estimate. Traditional approaches to eliminating such exchangeability is to impose identifiability constraints on the parameter space. However, as pointed out by Frühwirth-Schnatter [18], an unappropriate identifiability constraint may not be able to eliminate label switching. Many efforts have been devoted to coping with this issue, see Chapter 11 in Lee [47] for a review. Among them, the relabeling algorithm [48] is more appealing due to its simplicity and flexibility. The relabeling sampling procedure takes a decision-theoretical approach and requires specifying an appropriate loss function to measure the loss in terms of the classification probability. The model identification problem is addressed via postprocessing the MCMC output to minimize the posterior expected loss. Specifically, let $\boldsymbol{\theta}$ be the collection of $\boldsymbol{\Xi}$ and $\boldsymbol{\pi}$, and write $\mathbf{Q} = \{q_{ik}(\boldsymbol{\theta})$ as the matrix of allocation probabilities of order $n \times K$ with $q_{ik}(\theta) = \mathbb{P}(K_i = k|\mathbf{Y}, \ \boldsymbol{\theta})$. In the context of clustering, the loss function can be defined on the cluster label $\mathcal{K}$ as follows

$$\mathcal{L}_0(\mathcal{K}; \ \boldsymbol{\theta}) = -\sum_{i=1}^{n} \log q_{iK_i}(\boldsymbol{\theta}). \tag{10}$$

Given that $\boldsymbol{\theta}^{(1)}, \cdots, \ \boldsymbol{\theta}^{(M)}$ are the sampled parameters and let $\nu_1, \cdots, \nu_M$ be the permutation applied to them. The relabeling algorithm proceeds by selecting initial values for the ν_ms, which are generally taken to be the identity permutations, then iterating the following steps until a fixed point is reached.

a. Choose $\hat{\mathcal{K}}$ to minimize $\sum_{m=1}^{M} \mathcal{L}_0\left(\mathcal{K}, \nu_m\left(\boldsymbol{\theta}^{(m)}\right)\right)$;

b. For $m = 1, 2, \cdots, M$, choose ν_m to minimize $\mathcal{L}_0\left(\hat{\mathcal{K}}, \upsilon_m\left(\boldsymbol{\theta}^{(m)}\right)\right)$.

2.4 Posterior inference

Once the label switching is taken care of, the MCMC samples can be used to draw posterior inference. For example, the joint Bayesian estimate of $\boldsymbol{\theta}$ can be obtained easily via the corresponding sample means of the generated observations via ergodic average as follows:

$$\hat{\boldsymbol{\beta}}_k = M^{-1}\sum_{m=1}^{M} \boldsymbol{\beta}_k^{(m)}, \hat{\boldsymbol{\tau}}_k = M^{-1}\sum_{m=1}^{M} \boldsymbol{\tau}_k^{(m)}, \ \text{and}\, \hat{\pi}_k = M^{-1}\sum_{m=1}^{M} \pi_k^{(m)} \tag{11}$$

The consistent estimates of the covariance matrix of estimates can be obtained via sample covariance matrix.

Given the observations $\{\mathbf{K}^{(m)} : m = 1, 2, \cdots, M\}$ drawn from the posterior $p(\mathbf{K}|\mathbf{Y})$ via MCMC sampling, serval methods can be available for arriving at a point estimate of the clustering using draws from the posterior clustering distribution. The simplest method, known as the maximum a posteriori (MAP) clustering, is to select the observed clustering that maximizes the density of the posterior clustering distribution, i.e.,

$$\hat{\mathcal{K}} : \hat{K}_i = \text{argmax}_{k=1,\cdots,K} \mathbb{P}(K_i = k|\mathbf{Y}) \tag{12}$$

in which $\mathbb{P}(K_i = k|\mathbf{Y})$ can be approximated by

$$\mathbb{P}(K_i = k|\mathbf{Y}) \approx M^{-1} \sum_{m=1}^{M} I\left\{K_i^{(m)} = k\right\}. \tag{13}$$

A more appreciate alternative to MAP is based on the pairwise probability matrix, an $n \times n$ association matrix $\delta(\mathcal{K})$ with the (i,j)th element formed by the indicator of whether the subject i is clustered with subject j. Element-wise averaging of these association matrices yields the pairwise probability matrix of clustering, denoted $\hat{\psi}$. Medvedovic and Sivaganesan [49] and Medvedovic et al. [50] suggested a clustering estimate of $\mathcal{K}$ by using the pairwise probability matrix $\hat{\psi}$ as a distance matrix in hierarchical/agglomerative clustering. However, as augured by Dahl [51], such routine seems counterintuitive to apply an ad hoc clustering method on top of a model which itself produces clusterings. In the context of Dirichlet process mixture-based clustering, Dahl [51] proposed a least-squares model-based clustering method by using draws from a posterior clustering distribution. Specifically, the least-squares clustering $\mathcal{K}_{LS}$ is the observed clustering $\mathcal{K}_{LS}$, which minimizes the sum of squared deviations of its association matrix $(\mathcal{K})$ from the pairwise probability matrix:

$$\hat{\mathcal{K}}_{LS} = \text{argmin}_{\mathcal{K} \in \{\mathbf{K}^{(1)},\cdots,\mathbf{K}^{(m)}\}} \sum_{i=1}^{n} \sum_{j=1}^{n} (\delta(i,j)(\mathcal{K}) - \hat{\psi}(i,j))^2. \tag{14}$$

Dahl [51] showed that the least-squares clustering has the advantage over those in Medvedovic and Sivaganesan [49] since it utilizes the information from all the clusterings and is intuitively appealing for the "average" clustering instead of forming a clustering via an external, ad hoc clustering algorithm.

3. Assessing heterogeneity of two-part model

In this section, we first proposed a two-part regression model for the fractional data especially for the U shaped fractional data and then extend the method discussed above to the current situation to address the possible heterogeneity of the population underlying data.

3.1 Two-part model for U shaped fractional data

Suppose that for subject/individual $i(= 1, \cdots, n)$, y_i is an univariate fractional response taking values in $[0, 1]$; $\mathbf{x}_i$ is an $m \times 1$ fixed covariate vector denoting various explanatory factors under consideration. Usually, y_i suffers from excess zeros and ones on the boundaries, and the whole data set takes on the U shape. In modeling such data, we introduce a three-category indicator variable d_i and a continuous intensity variable z_i such that

$$d_i = \begin{cases} 1 & \text{if} \quad y_i = 0 \\ 2 & \text{if} \quad y_i = 1 \\ 3 & \text{if} \quad 0 < y_i < 1 \end{cases} \quad \text{and} \quad z_i = \begin{cases} h(y_i) & \text{if} \quad 0 < y_i < 1 \\ \text{irrelevant} & \text{if} \quad y_i = 0, 1 \end{cases} \tag{15}$$

where $h(\cdot)$ is any monotone increasing function such that $z_i \in (-\infty, +\infty)$. That is, we break the data set into three parts: two parts corresponding to zeros and ones respectively and one part corresponding to the continuous values between 0 and 1. We formulate a two-part model for y_i by first specifying a baseline-category logits model [52] for d_i and then a conditional continuous model for z_i. The baseline-category logits model is assumed that conditional upon $\mathbf{x}_i$, d_is are independent satisfying the following logits models simultaneously: for $j = 1, 2$,

$$\log \frac{\mathbb{P}(d_i = j|\mathbf{x}_i)}{\mathbb{P}(d_i = 3|\mathbf{x}_i)} = \mathbf{x}_i^T \boldsymbol{\alpha}_j \tag{16}$$

where $\boldsymbol{\alpha}_j$ is an $m \times 1$ regression coefficients vector. We use category $d_i = 3$ as the reference for the ease of parameters interpretation. For example, the magnitude of $\alpha_{j\ell}$ in $\boldsymbol{\alpha}_j$ indicates that the increase of one unit in $x_{i\ell}$ will increase $e^{\alpha_{j\ell}}$ times chance of $d_i = j$ over that of $d_i = 3$.

The conditional continuous model for z_i is given by

$$p(z_i|d_i = 3, \mathbf{x}_i) = p^z(z_i|\mathbf{x}_i^T \boldsymbol{\gamma}, \tau) \tag{17}$$

or equivalently

$$p(y_i|0 < y_i < 1, \mathbf{x}_i) = p^z(h(y_i)|\mathbf{x}_i^T \boldsymbol{\gamma}, \psi)|\dot{h}(y_i)| \tag{18}$$

where $\dot{h}(s) = dh/ds$, $p^z(u|a, \tau)$ is the normal density with mean a and variance $\tau > 0$, and $\boldsymbol{\gamma}$ like that in Eq. (16), is the regression coefficient vector. Although the identical covariates are taken in Eqs. (16) and (17), this is not necessary in practice. Each equation can own their covariates. This can be achieved by imposing particular structure on the regression coefficients. For example, we can exclude x_{i1} from Eq. (17) by restricting γ_1 in $\boldsymbol{\gamma}$ to be zero.

It follows from Eqs. (16) and (17) that marginal distribution of y_i is given by

$$p(y_i|\mathbf{x}_i, \boldsymbol{\beta}, \tau) = q_{i1}\delta_0 + q_{i2}\delta_1 + (1 - q_{i1} - q_{i2})p(y_i|0 < y_i < 1, \mathbf{x}_i, \ \boldsymbol{\gamma}, \tau) \tag{19}$$

where $q_{ij} = \mathbb{P}(d_i = j|\mathbf{x}_i, \ \boldsymbol{\alpha}_j)(\ j = 1, 2)$ is the response probability specified by Eq. (16) and $\boldsymbol{\beta}$ is the regression parameters constituted by $\boldsymbol{\alpha}_1, \boldsymbol{\alpha}_2$ and $\boldsymbol{\gamma}$.

3.2 Assessing heterogeneity of two-part model

To detect the possible heterogeneity among y_i, we extend the model Eq. (18) to the mixture case by assuming that conditional upon $K_i = k$, d_i and z_i satisfy Eqs. (16) and (17) with $\boldsymbol{\alpha}_j$ replaced by $\boldsymbol{\alpha}_{jk}$ and $(\boldsymbol{\gamma}, \tau)$ by $(\boldsymbol{\gamma}_k, \tau_k)$ respectively. This indicates that the mixture component f_k in Eq. (1) in Section 2 is given by Eq. (19) with $\boldsymbol{\beta} = \boldsymbol{\beta}_k$ and $\tau = \tau_k$.

For the Bayesian analysis, the general forms of full conditionals involved in the model-based clustering have been given in Section 2. We here only focus on the technical details of the conditional distribution of $\boldsymbol{\Xi}$ in (i) in the Gibbs sampler.

We assume that the prior of τ_k is the same as that in Eq. (6), while the priors of $\boldsymbol{\beta}_k$ are taken as $p(\boldsymbol{\beta}_k) = p(\boldsymbol{\alpha}_{k1})p(\boldsymbol{\alpha}_{k2})p(\boldsymbol{\gamma}_k)$, in which

$$p(\boldsymbol{\alpha}_{k\ell}) \stackrel{D}{=} N_m(\boldsymbol{\alpha}_{\ell 0}, \boldsymbol{\Sigma}_{\alpha\ell 0})(\ell = 1, 2), \quad p(\boldsymbol{\gamma}_k) \stackrel{D}{=} N_m(\boldsymbol{\gamma}_0, \boldsymbol{\Sigma}_{\gamma 0}). \tag{20}$$

where $\boldsymbol{\alpha}_{\ell 0}, \boldsymbol{\gamma}_0, \boldsymbol{\Sigma}_{\alpha\ell 0}$ and $\boldsymbol{\Sigma}_{\gamma 0}$ are the hyper-parameters treated to be known.

Gibbs sampling $\boldsymbol{\Xi}$ now becomes drawing $\boldsymbol{\alpha}_k, \boldsymbol{\gamma}_k$ and τ_k alternatively from the full conditional distributions $p(\boldsymbol{\alpha}_k|\mathbf{K}, \mathbf{Y}), p(\boldsymbol{\gamma}_k|\tau_k, \mathbf{K}, \mathbf{Y})$ and $p(\tau_k|\boldsymbol{\gamma}_k, \mathbf{K}, \mathbf{Y})$ respectively. By some algebras, it can be shown that

$$\begin{aligned} p(\boldsymbol{\alpha}_k|\mathbf{K}, \mathbf{Y}) &\propto p(\boldsymbol{\alpha}_k) \prod_{K_i=k} p(d_i|\mathbf{x}_i, \ \boldsymbol{\alpha}_k), \\ p(\boldsymbol{\gamma}_k|\tau_k, \mathbf{K}, \mathbf{Y}) &\propto p(\boldsymbol{\gamma}_k) \prod_{K_i=k} p\left(z_i|d_i = 3, \mathbf{x}_i^T \boldsymbol{\gamma}_k, \tau_k\right), \\ p(\tau_k|\boldsymbol{\gamma}_k, \mathbf{K}, \mathbf{Y}) &\propto p(\tau_k) \prod_{K_i=k} p\left(z_i|d_i = 3, \mathbf{x}_i^T \boldsymbol{\gamma}_k, \tau_k\right) \end{aligned} \tag{21}$$

in which the full conditionals of γ_k and τ_k are easily obtained and given by

$$p(\boldsymbol{\gamma}_k|\tau_k, \mathbf{K}, \mathbf{Y}) \stackrel{D}{=} N\left(\hat{\boldsymbol{\gamma}}_k, \hat{\boldsymbol{\Sigma}}_{\gamma k}\right) \tag{22}$$

$$p\left(\tau_k^{-1}|\boldsymbol{\gamma}_k, \mathbf{K}, \mathbf{Y}\right) \stackrel{D}{=} Gamma\left(\hat{\alpha}_k, \hat{\beta}_k\right) \tag{23}$$

in which

$$\begin{aligned} \hat{\boldsymbol{\Sigma}}_{\gamma k} &= \left(\sum_{K_i=k: d_i=2} \mathbf{x}_i \mathbf{x}_i^T / \tau_k + \boldsymbol{\Sigma}_{\gamma 0}^{-1} \right)^{-1}, \\ \hat{\boldsymbol{\gamma}}_k &= \hat{\boldsymbol{\Sigma}}_k \left(\boldsymbol{\Sigma}_{\gamma 0}^{-1} \ \boldsymbol{\gamma}_0 + \sum_{K_i=k, d_i=3} \mathbf{x}_i z_i / \tau_k \right), \\ \hat{\alpha}_k &= \alpha_0 + n_k/2, \\ \hat{\beta}_k &= \beta_0 + \sum_{K_i=k, d_i=3} \left(z_i - \mathbf{x}_i^T \boldsymbol{\gamma}_k\right)^2/2 \end{aligned} \tag{24}$$

and $n_k = \#\{K_i = k, d_i = 3\}$.

However, drawing $\boldsymbol{\alpha}_{k\ell}$ is more tedious since its distribution loses the standard form. We first note that

$$p(\boldsymbol{\alpha}_{k\ell}|\boldsymbol{\alpha}_{k,-\ell}, \mathbf{K}, \mathbf{Y}) \propto p(\boldsymbol{\alpha}_{k\ell}) \prod_{K_i=k}^{n} \frac{\exp\left(\tilde{d}_{i\ell}\left(\mathbf{x}_i^T \boldsymbol{\alpha}_{k\ell} - C_{ik\ell}\right)\right)}{1 + \exp\left(\mathbf{x}_i^T \boldsymbol{\alpha}_{k\ell} - C_{ik\ell}\right)} \tag{25}$$

where $\tilde{d}_{i\ell} = I\{d_i = \ell\}$ and $C_{ik\ell} = \log\left(1.0 + \exp\left(\mathbf{x}_i^T \boldsymbol{\alpha}_{k,-\ell}\right)\right)$; $\boldsymbol{\alpha}_{k,-\ell}$ denotes the set $\boldsymbol{\alpha}_k$ with $\boldsymbol{\alpha}_{k\ell}$ removed. Following the similar routine in Polson, Scott, and Windle [53], we recast the logistic function Eq. (25) as follows

$$\frac{\exp\left(\tilde{d}_{i\ell}\left(\mathbf{x}_i^T\boldsymbol{\alpha}_{k\ell}-C_{ik\ell}\right)\right)}{1+\exp\left(\mathbf{x}_i^T\boldsymbol{\alpha}_{k1}-C_{ik\ell}\right)}=2^{-1}\exp\left\{\kappa_{i1}\left(\mathbf{x}_i^T\boldsymbol{\alpha}_{k\ell}-C_{ik\ell}\right)\right\} \times\int_0^\infty\exp\left\{-\frac{1}{2}\omega_{i\ell}\left(\mathbf{x}_i^T\boldsymbol{\alpha}_{k\ell}-C_{ik\ell}\right)^2\right\}p_{\mathrm{PG}}(\omega_{i\ell})\,\mathrm{d}\omega_{i\ell} \tag{26}$$

in which $\kappa_{i\ell}=\tilde{d}_{i\ell}-1/2$ and p_{PG} is the well-known $\mathrm{PG}(1,0)$ density function [53]. If one introduces n independent Pólya-Gamma variables $\omega_{i\ell}$ into the current analysis, then,

$$p(\omega_{i\ell}|\,\boldsymbol{\alpha}_{k\ell},\boldsymbol{\alpha}_{k,-\ell},\mathbf{K},\mathbf{Y})\overset{D}{=}\mathrm{PG}\left(1,\left(\mathbf{x}_i^T\boldsymbol{\alpha}_{k\ell}-C_{ik\ell}\right)\right) \tag{27}$$

$$p(\boldsymbol{\alpha}_{k\ell}|\,\boldsymbol{\alpha}_{k,-\ell},\ \Omega,\mathbf{K},\mathbf{Y})\overset{D}{=}N\left(\hat{\boldsymbol{\alpha}}_{k\ell},\hat{\boldsymbol{\Sigma}}_{\alpha k\ell}\right) \tag{28}$$

where

$$\hat{\boldsymbol{\Sigma}}_{\alpha k\ell}=\left(\sum_{K_i=k}\mathbf{x}_i\mathbf{x}_i^T\omega_{i\ell}+\boldsymbol{\Sigma}_{\alpha\ell0}^{-1}\right)^{-1},\quad\hat{\boldsymbol{\alpha}}_{k\ell}=\hat{\boldsymbol{\Sigma}}_{\alpha k\ell}\left(\boldsymbol{\Sigma}_{\alpha\ell0}^{-1}\boldsymbol{\alpha}_{\ell0}+\sum_{K_i=k}\mathbf{x}_i\eta_{ik\ell}\right) \tag{29}$$

with $\eta_{ik\ell}=\kappa_{i\ell}+C_{ik\ell}\omega_{i\ell}$. Consequently, drawing $\boldsymbol{\alpha}_{k\ell}$ is accomplished by first drawing $\omega_{i\ell}$ from the Pólya gamma distribution and then drawing $\boldsymbol{\alpha}_{k\ell}$ from the normal distribution. The draw of $\omega_{i\ell}$ is a little intractable since its density function involves the infinite sum. By taking advantage of series sampling method [54], Polson et al. [53] devised a rejection algorithm for generating observations from such type of distribution. Their method can be adapted to draw $\omega_{i\ell}$, see also [55].

4. A real example

In this section, a small portion of cocaine use data is analyzed to illustrate the practical value of the proposed methodology. The data are obtained from the 322 cocaine use patients who were admitted in 1988–89 to the West Los Angeles Veterans Affairs Medical Center. The original data set is made up of 68 measurements in which 17 items were assessed at four unequally spanned time points. In this study, we mainly focus on the measurements 1 year after treatment and ignore the initial effects at the baseline. The measurements cover the information on the cocaine use, treatment received, psychological problems, social status, employments, and so on. Among them, the measurement "cocaine use per month" (denoted by CC) plays a critical role since it identifies the severity of cocaine use of patients and therefore is treated as the dependent response. The CC is originally measured by 0–30 points but suffered from small portion of fractions. We identify CC/30 as the fraction response in [0,1]. In view of that the missing data are presented, we delete the subjects with missing values. The total sample size is 228. A primary analysis shows that CC/30 has excessive zeros and ones. **Figure 2** gives the histograms of CC/30 and their fractional values in (0,1) via logistic transformation. It can be seen clearly that there is a large number of zeros and unities accumulated on the boundaries. The proportions of zeros and unities are about 15 and 4%, respectively. Moreover, panel (b) in **Figure 2** indicates that single parametric model may be unappreciate for fitting the continuous valued variable.

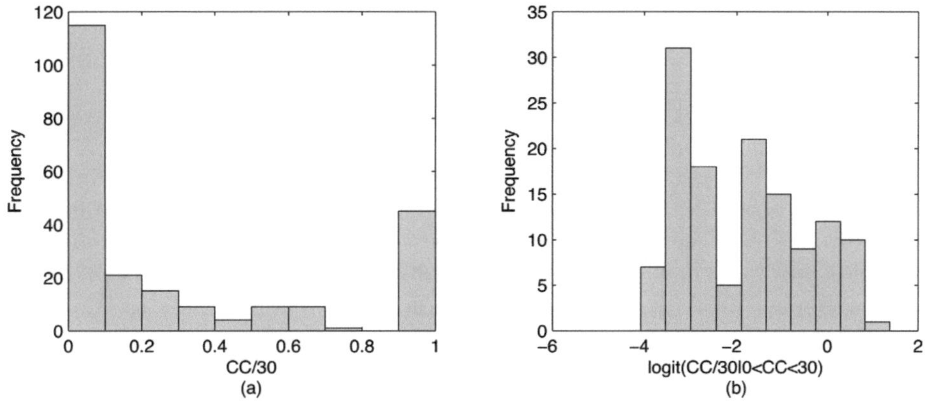

Figure 2.
Plots of CC in cocaine use data: (a) Histograms of CC/30; and (b) histograms of CC/30 on logistic transformation conditional on CC/30 in (0,1).

To explore the effects of exogenous factors on the cocaine use, the following measurements are selected as the explanatory variables: the occupational status of a patient (x_1). This is a binary indicator: 1 for employment and 0 for non-employment; the level of technical proficiency of patients engaged in work (x_2): scaled on 0–4 points and the patient's lifestyle (x_3) with five-point scale. To unify the scales, all covariates are standardized. However, a preliminary analysis shows that there exists strong multiple collinearity among these covariates. The minimum eigenvalue of sample covariance matrix equals to 0.06284, which approaches zero. We remove such collinearity by implementing principle component analysis (PCA) and treat the scores of the first two components (still denoted by x_1 and x_2) as our explanatory variables. These two principle components can be interpreted as the levels related to the patients' occupation and their live life.

To formulate a two-part model for the observed responses, we identity $CC_i/30$ with d_i and z_i, where d_i is the three-category indicator indicating the state of cocaine use after one year treatments: quitting cocaine successfully (state 1), insisting on cocaine use every day in a month (state 2) and taking the cocaine occasionally (state 3); z_i is the intensity variable representing the numer of days of cocaine use in a month. We assess the effects of exogenous factors x_1 and x_2 on the cocaine use via Eqs. (16) and (17), respectively.

We proceed data analysis by first fitting data to the K-component mixture two-part models with $K = 1, 2, \cdots, 6$. The model fits are assessed via AIC, AICc, and BIC, which are defined as $-2\log p\left(\mathbf{Y}|\hat{\boldsymbol{\theta}}_K\right)$ penalized by $2d_K$, $2n(d_K+1)/(n-d_K-2)$, and $d_K \log n$ respectively, where $\hat{\theta}_K$ is the MLE of $\boldsymbol{\theta}_K$ and d_K is the dimension of unknown parameters under the model K. In view of that the Bayesian estimates and the ML estimates are close to each other, we replace the ML estimates by their Bayesian counterparts in evaluating AIC, AICc, and BIC. For computation, we take $\alpha = n^{-1}, n^0, n^1$, and n^2 in Eq. (7), which represents our knowledge about $\boldsymbol{\pi}$ *a prior*. Note that for large value of α, the Dirichlet distribution places most of the mass on its center and the prior Eq. (7) tends to be informative. However, for small α, the Dirichlet distribution concentrates the mass on the boundaries of sampling space and the distribution tends to be degenerated and sparse. As a result, some components in $\boldsymbol{\pi}$ reduces to zeros. When $\alpha = 1$, $D_K(\alpha, \cdots, \alpha)$ becomes an uniform distribution on the simplex $\mathbb{S}_K$. For the inputs of the hyper-parameters involved in the priors Eq. (20), we take $\boldsymbol{\alpha}_{0\ell} = \boldsymbol{\gamma}_0 = \mathbf{0}_3$, $\boldsymbol{\Sigma}_{\alpha\ell 0} = \boldsymbol{\Sigma}_{\gamma 0} = 100\mathbf{I}_3$, $\alpha_{\gamma 0} = 2.0$ and $\beta_{\gamma 0} = 2.0$. These values ensure the priors Eq. (20) to be inflated enough and represent the weak information on the parameters.

The relabeling MCMC algorithm described in Section 2 is implemented to draw observations from the posterior. The convergence of algorithm is monitored by plotting the traces of estimates against iterations under three starting values. **Figure 3** presents the values of EPSR of unknown parameters against the number of iterations under three different starting values with $K = 2$. It shows that the convergence of the proposed algorithm is fast and the values of EPSR are less than 1.2 in less than 1000 iterations. Hence, 3000 observations, after removing the initial 2000 iterations, are collected for calculating AIC, AICc, and BIC. The resulting summary is given in **Table 1**.

Examination of **Table 1** shows that all measures favor the model with $K = 2$. This indicates that the proposed model with two groups seems to give a better fit to the data. It also indicates that large α favors the model fit. Furthermore, we calculate the posterior predictive density estimate of z_i under the elected model. Results (not represented here for saving spaces) show that our method can be successful in capturing the skewness and modes of data. We also follow [56] to plot the estimated residuals $\hat{\delta}_i = z_i - \hat{\boldsymbol{\gamma}}\mathbf{x}_i^T$ and find that these plots lie within two parallel horizontal lines that are centered at zero, with nonlinear or quadratic trends detected. This roughly indicates that the proposed linear model Eq. (18) is adequate.

Table 2 presents the estimates of unknown parameters associated with corresponding standard deviation (SD) estimates under $K = 2$. Based on **Table 2**, we can find the following facts: (i) for Part one, we observe that except for $\hat{\alpha}_{23}$, the Bayesian estimates of unknown parameters within two clusters have the same signs but their magnitudes are more different. For example, the estimate of α_{11} within Cluster one is -1.540 with SD 0.587 while equals to -0.732 with SD 0.481 within Cluster two. This indicates that the baselines of logits Eq. (16) exist obvious

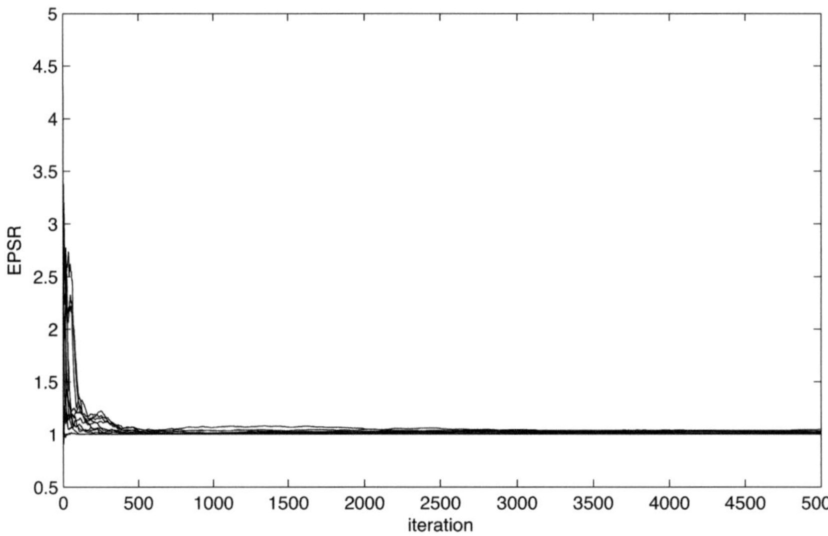

Figure 3.
Plots of values of EPSR of estimates of unknown parameters against the number of iterations under three different staring values in the cocaine use example: $K = 2$.

	Model	$\alpha = 1/n$	$\alpha = n^0$	$\alpha = n$	$\alpha = n^2$
AIC	$K = 1$	921.3887	–	–	–
	$K = 2$	923.0580	907.4485	901.9474	901.4380
	$K = 3$	929.0698	926.5423	956.4945	994.5039
	$K = 4$	990.7506	949.5966	1014.5477	1006.4228
	$K = 5$	989.2483	971.4688	1069.1561	1037.5005
	$K = 6$	1097.6853	1007.4899	1091.8049	1086.8491
AICc	$K = 1$	882.4025	–	–	–
	$K = 2$	885.5434	869.9339	864.4329	863.9234
	$K = 3$	875.9005	873.3730	903.3253	941.3347
	$K = 4$	925.3159	884.1618	949.1130	940.9880
	$K = 5$	915.5836	897.8041	995.4914	963.8357
	$K = 6$	1020.6483	930.4529	1014.7679	1009.8120
BIC	$K = 1$	995.6821	–	–	–
	$K = 2$	995.0742	979.4647	973.9637	973.4542
	$K = 3$	1038.8088	1036.2814	1066.2335	1104.2429
	$K = 4$	1138.2125	1097.0585	1162.0096	1153.8846
	$K = 5$	1174.4330	1156.6534	1254.3408	1222.6851
	$K = 6$	1320.5928	1230.3973	1314.7124	1309.7565

Table 1.
Summary statistics of AIC, AICcc, and BIC for model selection in cocaine use data analysis.

Para.	Component I		Component II	
	Est.	SD.	Est.	SD
α_{11}	−1.540	0.587	−0.732	0.481
α_{12}	0.150	0.317	0.604	0.322
α_{13}	0.261	0.703	0.188	0.601
α_{21}	−1.337	0.480	−1.059	0.545
α_{22}	−0.166	0.355	−0.229	0.418
α_{23}	0.232	0.378	−0.184	0.411
γ_1	−2.779	0.144	−0.490	0.215
γ_2	−0.029	0.080	−0.011	0.154
γ_3	0.087	0.144	0.179	0.240
τ	0.674	0.150	0.924	0.234

Table 2.
Summary statistics for the Bayesian estimates of unknown parameters in the cocaine use data.

difference. For α_{23}, the estimates between two clusters have the opposite signs. Recall that α_{23} quantifies the magnitude of effects of live life on the probability $\mathbb{P}(d_i = 2)$ over $\mathbb{P}(d_i = 3)$ on log scale. This shows that increasing the level of live life will lead to an opposite effect among two clusters; (ii) for Part two, although all the estimates within two clusters have the same signs but the levels of effects among them are obviously different. The estimates of γ_1 is -2.779 with SD 0.144 in the cluster one and attains -0.490 associated with SD 0.215 in the Cluster two. This indicates that the baseline of cocaine use in Cluster one is 50 times as much as that in Cluster two; and (iii) investigation of the estimate of τ also indicates that there exists the different amount of the fluctuation among two clusters.

5. Discussion

This chapter introduces a general Bayesian model-based clustering procedure for the regression model and proposed a Bayesian method for assessing the heterogeneity of fractional data within the mixture of two-part regression model framework. The heterogeneous fractional data arise mainly from two resources: one is that the excessive zeros and ones are accumulated upon the boundaries, and the other is that the underlying population may consist more than one components. For the first issue, we propose a novel two-part model, in which a three-category multinomial regression is suggested to model the occurrence probabilities of each part, and a conditional normal linear regression is used to fit the continuous positive values on logit scale. Such formulation is more appealing since it can ensure the probabilities on each part to be consistent and and at the same time maintains the coherent association across parts. For the second problem, we resort to the finite mixture model in which the cluster-specific components are specified via two-part model. MCMC sampling method is adopted to carry out the posterior analysis. The number of clusters and the configuration of observations are addressed based on the simulated observations from the posterior. We illustrate the proposed methodology in the analysis of cocaine use data.

When interest is concentrated upon the estimates, model identification is surely an important issue since it involves whether or not the estimates of component-specific quantities are meaningful. For a finite mixture model, model identification mainly stems from the label switching, in which the likelihood and the posterior are invariant under label permutation. Many efforts have devoted to alleviating such indeterminacy. Among them, parameters' constraints may be the most popular choice. However, an unappreciated constraint fails to deal with the label switching. In this case, one can follow the routine in Frühwirth-Schnatter [18] and implement random permutation sampling to find the suitable identifiability constraints. The random permutation sampler is similar to the unconstrained MCMC sampling but only at each sweep, the labels $\{1, \cdots, K\}$ are randomly permutated. The permutation aims to deliver a sample that explores the whole unconstrained parameter space and jumps between the various labeling subspaces in a balanced fashion. The output of such balanced sample can help us to find a suitable identifiability constraint. A more detailed discussion on model identification in the mixture context can be referred to, for example, [18, 57]. Instead, we resort to the relabeling algorithm for simplicity. Compared with the random permutation sampling, the relabeling method requires implementing MCMC samplng only once, thus saving the computation cost.

The methodology developed in this chapter can be extended to the case where latent factors are included to identify the unobserved heterogeneity due to some fixed convariates absent. Another possible extension is to establish a dynamic LVM, wherein model parameters vary across times. These issues may raise theoretical and computational challenges and therefore require further investigation.

Acknowledgements

The work presented here was fully supported by grant from the National Natural Science Foundation of China (NNSF 11471161). The authors are thankful to Professor Xin-Yuan Song, the Chinese University of Hong Kong, for using her cocaine use data in the real example.

Conflict of interest

The authors have no conflicts of interest to disclose.

Author details

Ye-Mao Xia[1], Qi-Hang Zhu[2] and Jian-Wei Gou[1]*

1 Department of Applied Mathematics, Nanjing Forestry University, Nanjing, China

2 College of Economics and Management, Nanjing Forestry University, Nanjing, China

*Address all correspondence to: gjw1983@139.com

References

[1] McLachlan GJ. Discriminant Analysis and Statistical Pattern Recognition. New York: John Wiley; 1992. DOI: 10.1002/0471725293.ch3

[2] Fraley C, Raftery AE. Model-based clustering, discriminant analysis, and density estimation. Journal of the American Statistical Association. 2002; **97**(458):611-631. DOI: 10.2307/3085676

[3] Andrews JL, McNicholas PD. Model-based clustering, classification, and discriminant analysis via mixtures of multivariate t-distributions: The tEIGEN family. Statistics and Computing. 2012; **22**(5):1021-1029. DOI: 10.1007/s11222-011-9272-x

[4] Ripley BD. Pattern Recognition and Neural Networks. Cambridge, UK: Cambridge Univeristy Press; 1996. DOI: 10.1080/00401706.1997.10485099

[5] Paalanen P, Kamarainen JK, Ilonen J, Kälviäinen H. Feature representation and discrimination based on Gaussian mixture model probability densities Practices and algorithms. Pattern Recognition. 2006;**39**(7):1346-1358. DOI: 10.1016/j.patcog.2006.01.005

[6] Qin LX, Self SG. The clustering of regression models method with applications in gene expression data. Biometrics. 2006;**62**:526-533

[7] McNicholas PD, Murphy TB. Model-based clustering of microarray expression data via latent Gaussian mixture models. Bioinformatics. 2010; **21**:2705-2712. DOI: 10.1093/bioinformatics/btq498

[8] Yuan M, Kendziorski C. A unified approach for simultaneous gene clustering and differential expression identification. Biometrics. 2006;**62**:1089-1098

[9] Kanungo T, Mount DM, Netanyahu NS, Piatko CD, Silverman R, Wu AY. An efficient k-means clustering algorithm: analysis and implementation. IEEE Transactions on Pattern Analysis & Machine Intelligence. 2002;**24**(7): 881-892. DOI: 10.1109/TPAMI.2002.1017616

[10] Mahmoudi MR, Akbarzadeh H, Parvin H, Nejatian S, Alinejad-Rokny H. Consensus function based on cluster-wise two level clustering. Artificial Intelligence Review. 2021;**54**:639-665. DOI: 10.1007/s10462-020-09862-1

[11] MacQueen J. Some methods for classification and analysis of multivariate observations. In: Cam LML, Neyman J, editors. Proceedings of the 5th Berkeley Symposium on Mathematical Statistics & Probability. Vol. 1. Berkeley, CA: University of California Press; 1967. pp. 281-297

[12] Hartigan JA, Wong MA. Algorithm AS 136: A K-means clustering algorithm. Journal of the Royal Statistical Society, Series C. 1979;**28**(1):100-108. DOI: 10.2307/2346830

[13] Anderberg MR. Cluster Analysis for Applications. New York: Academic Press; 1973

[14] Everitt BS, Landau S, Leese M. Cluster Analysis. 4th ed. London: Hodder Arnold; 2001

[15] Johnson RA, Wichern DW. Applied Multivariate Statistical Analysis. 2th ed. New Jersey: Prentice Hall; 1988

[16] Titterington DM, Smith AFM, Makov UE. Statistical Analysis of Finite Mixture Distributions. Chichester: John Wiley and Sons; 1985. DOI: 10.2307/2531224

[17] McLachlan GJ, Peel D. Finite Mixture Models. New York: John Wiley; 2000. DOI: 10.1002/0471721182

[18] Frühwirth-Schnatter S. Markov chain monte carlo estimation of classical and dynamic switching and mixture models. Journal of the American Statistical Association. 2001, 2001; **96**(453):194-209. DOI: 10.1198/016214501750333063

[19] Fang KN, Ma SG. Three-part model for fractional response variables with application to Chinese household health insurance coverage. Journal of Applied Statistics. 2013;**40**(5):925-940. DOI: 10.1080/02664763.2012.758246

[20] McCullagh P, Nelder JA. Generalized Lincar Modcls. London: Chapman and Hall; 1989. DOI: 10.1007/978-1-4899-3242-6

[21] Green PJ. Reversible jump Markov chain Monte Carlo computation and Bayesian model determination. Biometrika. 1995;**82**(71):17-32. DOI: 10.1093/biomet/82.4.711

[22] Richardson S, Green PJ. On Bayesian analysis of mixtures with an unknown number of components (with discussion). Journal of the Royal Statistical Society, Series B. 1997;**59**:731C792. DOI: 10.1111/1467-9868.00095

[23] Akaike H. Information theory and an extension of the maximum likelihood principle. In: Petrov BN, Csáki F, editors. Second International Symposium on Information Theory. Budapest, Hungary: Akad¨¦mia Kiad¨®; 1973. pp. 267-281. DOI: DOI.10.1007/978-1-4612-1694-0-15

[24] Sugiura N. Further analysis of the data by Akaikes information criterion and the finite corrections. Communications in Statistics-Theory and Methods. 1978;**A7**:13-26

[25] Hurvich CM, Tsai C-L. Regression and time series model selection in small samples. Biometrika. 1989;**76**:297-307. DOI: 10.1093/biomet/76.2.297

[26] Schwarz G. Estimating the dimension of a model. The Annals of Statistics. 1978;**6**:461-464. DOI: 10.1214/aos/1176344136

[27] Biernacki C, Celeux G, Govaert G. Assessing a mixture model for clustering with the integrated completed likelihood. IEEE Transactions on Pattern Analysis and Machine Intelligence. 2000;**22**(7):719-725. DOI: 10.1109/34.865189

[28] Berger JO. Statistical Decision Theory and Bayesian Analysis. New York: Springer-Verlag; 1985. DOI: 10.1007/978-1-4757-4286-2

[29] Kass RE, Raftery AE. Bayes factors. Journal of the American Statistical Association. 1995;**90**:773-795. DOI: 10.1080/01621459.1995.10476572

[30] Spiegelhalter DJ, Best N, Carlin B, van der Linde A. Bayesian measures of model complexity and fit. Journal of the Royal Statistical Society, Series B. 2002; **64**:583-640. DOI: 10.1111/1467-9868.00353

[31] Spiegelhalter DJ, Thomas A, Best NG, Lunn D. WinBUGS User Manual. Version 1.4. Cambridge, England: MRC Biostatistics Unit; 2003. DOI: 10.1001/jama.284.24.3187

[32] Ferguson TS. A Bayesian analysis of some nonparametric problems. The Annals of Statistics. 1973;**1**(2):209-230. DOI: 10.1214/aos/1176342360

[33] Antoniak CE. Mixtures of Dirichlet processes with applications to bayesian nonparametric problems. The Annals of Statistics. 1974;**2**:1152-1174. DOI: 10.1214/aos/1176342871

[34] Ishwaran H, James LF. Gibbs sampling methods for stickbreaking priors. Journal of the American Statistical Association. 2001;**96**: 161-173. DOI: 10.1198/016214501750332758

[35] Dempster A, Laird N, Rubin D. Maximum likelihood from incomplete data via the EM algorithm (with discussion). Journal of the Royal Statistical Society, Series B. 1977;**39**:1-38

[36] Diebolt J, Robert CP. Estimation of finite mixture distributions through Bayesian sampling. Journal of the Royal Statistical Society, Series B. 1994;**56**: 363-375. DOI: 10.1111/j.2517-6161.1994.tb01985.x

[37] Roeder K, Wasserman L. Practical Bayesian density estimation using mixtures of normals. Journal of the American Statistical Association. 1997; **92**:894-902. DOI: 10.1080/01621459.1997.10474044

[38] Geman S, Geman D. Stochastic relaxation, Gibbs distribution, and the Bayesian restoration of images. IEEE Transactions on Pattern Analysis and Machine Intelligence. 1984;**6**:721-741. DOI: 10.1109/TPAMI.1984.4767596

[39] Geyer CJ. Practical Markov chain Monte Carlo. Statistical Science. 1992;7: 473-511. DOI: 10.1214/ss/1177011137

[40] Tanner MA, Wong WH. The calculation of posterior distributions by data augmentation(with discussion). Journal of the American statistical Association. 1987;**82**:528-550. DOI: 10.2307/2289463

[41] Gelfand AE, Smith AFM. Sampling-based approaches to calculating marginal densities. Journal of the American Statistical Association. 1990;**85**:398-409. DOI: 10.1080/01621459.1990.10476213

[42] Ishwaran H, Zarepour M. Markov chain Monte Carlo in approximation Dirichlet and beta-parameter process hierarchical models. Biometrika. 2000; **87**:371-390

[43] Gelman A, Rubin DB. Inference from iterative simulation using multiple sequences. Statistical Science. 1992;**7**: 457-472. DOI: 10.2307/2246093

[44] Metropolis N, Rosenbluth AW, Rosenbluth MN, Teller AH, Teller E. Equations of state calculations by fast computing machines. Journal of Chemical Physics. 1953;**21**:1087-1092. DOI: 10.1063/1.1699114

[45] Hastings WK. Monte Carlo sampling methods using Markov chains and their applications. Biometrika. 1970;**57**(1): 97-109. DOI: 10.1093/biomet/57.1.97

[46] Gilks WR, Wild P. Adaptive rejection sampling for gibbs sampling. Journal of the Royal Statistical Society. Series C (Applied Statistics). 1992;**41**(2): 337-348. DOI: 10.2307/2347565

[47] Lee SY. Structural Equation Modeling: A Bayesian Approach. New York: John Wiley & Sons; 2007

[48] Stephens M. Dealing with label-switching in mixture models. Journal of the Royal Statistical Society, Series B. 2000;**62**:795-809. DOI: 10.1111/1467-9868.00265

[49] Medvedovic M, Sivaganesan S. Bayesian infinite mixture model based clustering of gene expression profiles. Bioinformatics. 2002;**18**(9): 1194-1206. DOI: 10.1093/bioinformatics/18.9.1194

[50] Medvedovic M, Yeung KY, Bumgarner RE. Bayesian mixture model based clustering of replicated microarray data. Bioinformatics. 2004;**20**(8):

1222-1232. DOI: 10.1093/bioinformatics/bth068

[51] Dahl DB. Model-based clustering for expression data via a Dirichlet process mixture model. In: Do KA, Müller P, Vannucci M, editors. Bayesian Inference for Gene Expression and Proteomics. Cambridge University Press; 2006. DOI: 10.1017/CBO9780511584589.011

[52] Agresti A. Categorical Data Analysis. 2nd ed. New York: John Wiley & Sons; 2003

[53] Polson NG, Scott JG, Windle J. Bayesian inference for logistic models using pólya Gamma latent variables. Journal of the American Statistical Association. 2013, 2013;**108**(504): 1339-1349. DOI: 10.1080/01621459.2013.829001

[54] Devroye L. The series method in random variate generation and its application to the Kolmogorov-Smirnov distribution. American Journal of Mathematical and Management Sciences. 1981;**1**:359-379. DOI: 10.1080/01966324.1981.10737080

[55] Gou JW, Xia YM, Jiang DP. Bayesian analysis of two-part nonlinear latent variable model: Semiparametric method. Statistical Modeling. Published on line. 2021. DOI: 10.1177/1471082X211059233

[56] Xia YM, Tang NS, Gou JW. Generalized linear latent models for multivariate longitudinal measurements mixed with hidden Markov models. Journal of Multivariate Analysis. 2016;**152**: 259-275. DOI: 10.1016/j.jmva.2016.09.001

[57] Jasra A, Holmes CC, Stephens DA. Markov Chain Monte Carlo methods and the label switching problem in Bayesian mixture modeling. Statistical Science. 2005;**20**(1):50-67. DOI: 10.1214/088342305000000016

Section 3

Application of Clustering Methods

Chapter 5

Application of Jump Diffusion Models in Insurance Claim Estimation

Leonard Mushunje, Chiedza Elvina Mashiri, Edina Chandiwana and Maxwell Mashasha

Abstract

We investigated if general insurance claims are normal or rare events through systematic, discontinuous or sporadic jumps of the Brownian motion approach and Poisson processes. Using firm quarterly data from March 2010 to December 2018, we hypothesized that claims with high positive (negative) slopes are more likely to have large positive (negative) jumps in the future. As such, we expected salient properties of volatile jumps on the written products/contracts. We found that insurance claims for general insurance quoted products cease to be normal. There exist at times some jumps, especially during holidays and weekends. Such jumps are not healthy to the capital structures of firms, as such they need attention. However, it should be noted that gaps or jumps (unless of specific forms) cannot be hedged by employing internal dynamic adjustments. This means that, jump risk is non-diversifiable and such jumps should be given more attention.

Keywords: insurance claims, jumps, diffusion models, insurance claims, general insurance, volatility, reserving

1. Introduction

Insurance claim jumps are irregularities of the claims frequency from the policyholder to the insurer. They are crucial and fundamental in the understanding and tackling of insurable risks. We therefore, explore insurance claim jumps in general insurance products. Specifically, we investigate whether claims are associated with systematic, sporadic or discontinuous jumps or they undergo through a normal process. Our aim was to explore if insurance claims are rare or normal events using the Brownian motion model and Poisson processes in testing diffusion and jump risk. The second aim was to explore how the identified jumps affect the company's solvency status. We put forward that, the knowledge of claim jumps is useful in proper pricing of products and better claim reserve calculations. We hypothetically state that, persistent claim jumps lead to the ruin problem. Furthermore, we conjectured that claims with high positive (negative) slopes are more likely to have large positive (negative) jumps in the future. A mismatch between liabilities and assets is central to insurance.

High frequency claims exhibit fat-tailed distributions (excess kurtosis), skewness and are in most cases clustered together. We can say, the infrequent movements of large magnitude in claim counts are attributed to sudden-jumps that we want to really explore in this study.

Literally, diffusion models are tools used to describe the movement, decay and evolution of products or items in a given environment over a specified period. The variables are normally random in nature. The general application of diffusion processes is to describe the evolution of asset/product's behavior over time in terms of their prices or returns. In finance, we see the application of the models in explaining the evolution of asset returns [1, 2]. Randomness and persistence are two salient properties of claim jumps and volatility. Jump diffusion models are applied in the financial arena to estimate stock volatilities of both prices and returns [3–6]. The statistical properties of claim amounts have long been of curiosity to insurers and actuaries in pricing and risk management. Higher order expectations are less considered as much of the information about any financial or insurance data is believed to have been carried by the standard arithmetic mean and standard deviations. However special cases like positive kurtosis implies concave, U-shaped implied Black-Scholes volatility (IV) curves. Practitioners rarely do taking higher order expectations in statistical distributions such as the Gaussian, Binomial, and Poison and so on.

Claimant distributions are Longley modeled using the compound poison and gamma distributions where the former captures the frequency and the latter captures the severity of the claims [7]. No serious attention was exerted to the jumps associated with the insurance claims and the jump effects too. In general, insurance where the frequency of claim arrival is high, jump analysis is quite necessary utmost for economical reserving and capital solvency.

In option pricing, the use of Black and Scholes-type formulae is considered to price European options on written underlying assets such as stocks, foreign currencies, commodities and interest rates. We therefore intend to apply the taste of diffusion models in modeling the evolution and behavior of insurance claims for the written non-life products. The Jump Diffusion model chosen in this study can potentially explain the evolution of claims and its behavior (frequency and severity) more accurately at the expense of making the market incomplete, since jumps in premiums cannot be hedged easily. The reason behind the existence of claim reserves such as the unearned premium reserves is a key indication of the jumps in premium payments. However, underestimation is commonly burdening insurers. Underestimation is a tendency of deriving and providing values, which are excessively low and unfavorable. In our context, underestimation negatively affects the insurer's capital structures and reserving. Thus, jump is indeed an important aspect that should be taken into consideration at regular times. We do this using the Gaussian, Poison model and by extending Merton's [1, 8] jump-diffusion model, which we presented in our methods section. The remainder of our paper is organized as follows. The next section generalizes our jump diffusion models to a firm level. Empirical tests for the presence of jump components in the claims are contained in Section 3. Section 4 concludes the paper.

2. Materials and tools

The study employed the model contained in [1, 5]. We interpret the model as the one which contains a finite number of insurance contracts and insurers and insured. The model is based on the following assumptions:

1. No transaction costs, no taxes, and frictionless insurance markets.

2. Competitive markets (insurers are price (premiums) takers)

3. Continuous trading at equilibrium prices (premiums)

4. There are m risky contracts whose premiums and claims satiate

$$\frac{dC_j}{C_j} = \alpha_j dt + \sigma_j dZ_j + \left(-\lambda_j K_j dt + \pi_j dY_j\right), j = 1 : m \dots. \quad (1)$$

where $C_j(t)$ is the claim amount of a contract j at time t; α_j, σ_j, λ_j, and K_j are constants where α_j and σ_j are the drift and diffusion components respectively; dZ_j is a Wiener process; dY_j is a Poisson process with parameter λ_j; π_j is the jump amplitude with expected value equal to K_j; and dZ_j, dY_j, and π_j are independent.

5. Further, Insurers have standardized opinions over $\{\alpha_j, \sigma_j, \lambda_j, K_j$, j = 1 ... m$\}$

6. Insurers' and insured's tastes are represented by a von Neumann-Morgenstern utility functional theory which is strictly increasing and strictly concave. All our assumptions except 4 are found commonly in literature [1, 5]. Assumption number 4 is the key conjecture in our analysis.

We now rewrite assumption 4 in an equivalently alternate way that separates systematic and unsystematic risk components.

Consider the diffusion part of assumption 4,

$$dD_j = \alpha_j dt + \sigma_j dZ_j; j = 1; \dots; m \dots. \quad (2)$$

Following the argument from Ross [9], expression (2) implies that there exists. $\{u_j, f_j, g_j, d\varnothing, dW_j\}$; j = 1, ... m, such that

$$dD_j = \alpha_j dt + f_j d\varnothing + g_j W d_j; j = 1; \dots; m \dots. \quad (3)$$

where $fj_2 + g_j^2 = \sigma_j^2$; $d\varnothing$, dW_j are Wiener processes;

$$E\left[d\varnothing dW_j\right] = 0, j = 1, \dots, m;$$

and,

$$\sum_{j=1}^{m} u_j = 1, (x+a)^n = \sum_{j=1}^{m} u_j\left(g_j dW_j\right) = 0 = \sum_{j=1}^{m} u_j \alpha_j > r \dots \quad (4)$$

It is always likely to decompose a restricted number of normal arbitrary variables into a common factor, $d\varnothing$, and error terms, dW_j, which are normally distributed. The key property of normal claims employed is that covariance of zero implies numerical independence. This same assumption is confirmed in asset prices and returns analysis [10]. Note that $d\varnothing$, dW_j will be independent of dY_j and π_j by assumption 4. This disintegration gives $d\varnothing$ the interpretation of being the unsystematic risk factor.

Substitution of expression (3) and (4) into (1) gives assumption 6: There are m risky insurance contracts whose claims satisfy:

$$\frac{dC_j}{C_j} = \alpha_j dt + f_j d\emptyset + g_j dW_j + (-\lambda_j K_j dt + \pi_j dY_j), j = 1, \ldots, m \ldots \quad (5)$$

where $C_j(t)$ is the claim of a contract j at time t; $\alpha_j, f_j, g_j, \lambda_j, K_j$ are constants; $d\emptyset$, dW_j are Wiener processes; dY_j is a Poisson process with parameter λ_j; π_j is the jump amplitude with expected value equal to K_j; and $d\emptyset$, dW_j, dY_j, π_j are independent. The jump component in expression (5), $(-\lambda_j K_j dt + \pi_j dY_j)$, infers that insurance claims can have discontinuous ample paths. This generalizes existing models.

3. Data and model

The section tests the written insurance contracts claims to see if they contain jumps. If no jump component is present, then this would be consistent with the proposition of the previous deduction. In addition, it implies that the claims are normal events. Thus, the satisfaction of instantaneous claim reserves calculation frameworks such as the Chain ladder method and pricing models (collective risk model). We used the written insurance contacts and the recorded claims for a period spanning from March 2010 to December 2018. We performed the following hypothesis tests:

H_0, jump risk is diversifiable.

H_1, jump risk is non-diversifiable.

From the above hypothesis, we will see whether jump risk leads to capital insolvency for insurance firms. We will survey the sample path of the claims. To advance the testing procedure, note that under expression (5) the insurance claims dynamics are given by:

$$\frac{dC}{C} = \sum_{j=1}^{m} C_j \alpha_j dt + \left(\sum_{j=1}^{m} C_j f_j\right) d\emptyset + \sum_{j=1}^{m} C_j \left(g_j dW_j - \lambda_j K_j dt \pi_j dY_j\right) + log\ V_j \quad (6)$$

Where, $C = \sum_{j=1}^{n} m_j C_j$, $log\ V_j \sim i.i.d.N(\alpha, \sigma^2)$, normally distributed and models jumps in claims. Under the null hypothesis, expression (6) reduces to:

$$\frac{dC}{C} = \alpha dt + \sigma d\emptyset \ldots \quad (7)$$

Where, $\alpha = \sum_{j=1}^{n} m_j \alpha_j$ and $\sigma = \sum_{j=1}^{n} m_j f_{j.}$

Under the alternative hypothesis, expression (6) reduces to:

$$\frac{dC}{C} = \alpha' dt + \sigma d\emptyset + dq \ldots \quad (8)$$

where $dq = \pi dY$ denotes a Poisson process with parameter λ, $\pi =$ jump amplitude with estimated value equal to K, and $\alpha' = \alpha — \lambda K$.

Another assumption is added to (8), that is, (π) has a lognormal distribution with parameters (a, b^2). We add this assumption to easy up the Maximum Likelihood Estimation procedure in estimating the parameters of Eqs. (7) and (8).

Component	Statistic value
Constant drift	0.21%
Drift deviance	2%
Probability jumps	4%
Mean of jumps	3%
Jump deviance	5%

Table 1.
Diffusion process parameter estimates.

Now, we conveniently re-write the hypothesis to be tested as follows:
H_0, jump risk is diversifiable

$$\frac{dC}{C} = \alpha dt + \sigma d\emptyset \dots . \tag{9}$$

H_1, jump risk is non-diversifiable

$$\frac{dC}{C} = \alpha' dt + \sigma d\emptyset + dq \dots . \tag{10}$$

and (π) is dispersed lognormal (a, b^2)

Now, to properly test the above stated null hypothesis, a likelihood ratio test can be used: $A = -2(\ln L_r — \ln L_u)$, where L_r is the likelihood value for the reserved density function (i.e., the null hypothesis, Eq. (9)) and L_u signifies the likelihood function for the unconstrained density function (i.e., the alternative hypothesis, Eq. (10)).

Table 1 presents estimates of parameters of the diffusion-only process for diverse observation intervals and time periods. The results suggest that the total claims frequency and severity are not constant over time. The total standard deviation of claims on the firm is measured by the total claims index over 8-year period.

Table 1 is a summary of the parameter estimates of the diffusion model used over a time horizon for a basket of diversified observations. Having the parameter values the jump-diffusion model can be safely used to infer the likely consequences of the claim jumps towards an efficient insurance engineering. The jump probability is our spanner for dealing with ruin issues and proper reserve estimations. Jump deviance is the standard deviation of the jumps, which gives the spread of the claim jumps (positive or negative) over time for the written contracts.

4. Methodology

Throughout this paper we assume that C_t to be the claim amount of each insurance contract at time t, whose dynamics are given by;

$$\frac{dC_t}{C_t} = (\mu - \lambda\kappa)dt + \sigma dBt + \left[e^J - 1\right]dNt, , , , , , \tag{11}$$

where μ, is the instantaneous expected claim amount per unit time, and σ is the instantaneous volatility per unit time. The stochastic process B_t is a standard Wiener process under the market measure P. The process $N(t)$ is a Poisson process, independent of the jump-sizes J and the Wiener process Bt, with arrival intensity λ per unit time under the measure P, so that its increments satisfy the following:

$$d(Nt) = \begin{Bmatrix} 1,,,,,,,with\ probability\ (\lambda dt) \\ 0,,with\ probability\ (1-\lambda dt) \end{Bmatrix},,,,,, \tag{12}$$

The expected proportional jump size is;

$$\kappa = E\left[e^J - 1\right],,,,,,. \tag{13}$$

In this study, jumps are assumed independent of each other as they arrive at different times. We then defined an information set through a filtered probability measure space $(\Omega, F, \{Ft\}, P)$, where the filtration $\{Ft\}$ is the natural filtration generated by the Wiener process Bt. In the jump-diffusion model, the insurance claims C_t are defined to follow the random process given by:

$$\frac{dC_t}{C_t} = \mu dt + \sigma dW_t + (J-1)dN_t,,,,,. \tag{14}$$

The first two terms are familiar from the Black Scholes [11] model: The drift rate μ, volatility σ, and random walk (Wiener process), Wt. The last term represents the jumps: J is the jump size and $N(t)$ is the number of jump events that have occurred up to time t. $N(t)$ is assumed to follow the Poisson process;

$$P((Nt) = k) = \frac{(\lambda t)^k}{k!} e^{-\lambda t},,,,, \tag{15}$$

where, λt is the average number of jumps per unit time. Note that, there is no specific distribution for the jump sizes. However, a common choice is a log-normal distribution given as:

$$J \sim me^{\frac{v^2}{2} + vN(0,1)},,,,, \tag{16}$$

where $N\ (0,1)$ is the standard normal distribution, m is the average jump size, and v is the volatility of jump size. The key parameters that characterize the jump-diffusion model are λ, m, v.

5. Model

We use the basic excel spreadsheet to model the effects of jump-diffusion on claims and the respective reserves. Our equation is as follows:

$$r_t = \alpha + \varepsilon_t + I_t u_t,,,,, \tag{17}$$

where r_t is the log claim amount, α is the mean drift, ε_t is the diffusion which follows a normal distribution calculated as $\sigma * NORMSINV(RAND())$, where σ is the standard deviation of the jumps, I is the indicator variable (0 or 1), for either absence or presence of the claim jump. The value is determined by the jump probability; u_t is the value of the jump. This follows a normal distribution and is determined by

$$E[u] + \sigma_u * NORMSINV(RAN()),$$

where $E[u]$ and σ_u are the mean and standard deviation of the jump, respectively. Space is too much

6. Results

Using Excel Visual Basic for Applications (VBA) and R scrip, we perform our analysis, using the calibrated parameters. Parameter calibration was done using the maximum likelihood approach. Concisely, we presented the model results in a nicely and user-friendly manner. The user can only enter the input values on the designed user form and click 'run'. The inputs included are the sigma (volatility value that we normally called the implied volatility), the risk free interest rate, time component (T), the number of paths for simulations (we used 174 for our case, but can be varied). **Table 2** summary is in the subsequent tables in the Appendices section. The number of jumps are then estimated and modeled within the selected paths number and period. Ones denote jumps, otherwise they are normal claim movements.

7. Discussions

We tested whether or not there are systematic jumps insurance claims or they are normal events. We found that insurance claims for general insurance quoted products cease to be normal. There exist at times some jumps, especially during holidays and weekends. Such jumps are not healthy to the capital structures of firms, as such, they need attention. However, it should be noted that gaps or jumps (unless of specific forms) cannot be hedged by employing internal dynamic adjustments. This means that, according to our hypothesis tested, jump risk is non-diversifiable. A firm can manage jump-induced risks by buying options. Option derivatives help the firm to protect it against negative jumps and its consequences on its capital status. If, however, it establishes its own reserves, it must ensure and enforce a dynamic reserve adjustment. The reserves must increase as position values fall. This is an alternative option. The insurers must bear in mind the cushion, so to speak, that is dynamic. However, by dynamically hedging its own capital account, the insurer cannot wholly protect itself. Gaps or jumps are truly difficult to hedge; we thus need an idea of option hedging.

8. Conclusion

This paper develops and tests sufficient conditions for a model when insurance claims follow a jump-diffusion process. Based on weekly claims data, our results are that the reported claims contain a jump component, with a slightly high magnitude. We measure the jump component over both short (monthly and larger intervals in time (quarterly interval) and find that the weekends and holidays tend to cover up the high jump component. The economic intuition is that jump risk is not diversifiable and hence can ruin the firm leading to capital insolvency.

Appendix

Run Button

Time period (t)	Diffusion (ε)	Jump?	Jump value (u)	log (claims)	log (premium)	Jump (%)
0					0.661	
1	0.0003	0.0000	0.0121	0.1204	0.6620	
2	0.0003	0.0000	0.0135	0.1215	0.6610	
3	0.0005	0.0000	0.0120	0.1311	0.6600	
4	0.0003	0.0000	0.0245	0.1216	0.6600	
5	0.0007	0.0000	0.0245	0.1478	0.6580	
6	0.0003	0.0000	0.0245	0.1207	0.6570	
7	0.0016	0.0000	0.0245	0.3174	0.6580	
8	0.0016	0.0000	0.0126	0.3174	0.6560	
9	0.0016	0.0000	0.0245	0.3174	0.6560	
10	0.0016	0.0000	0.0146	0.3174	0.6540	
11	0.0005	0.0000	0.0267	0.1317	0.6540	
12	0.0016	0.0000	0.0256	0.3174	0.6540	
13	0.0011	0.0000	0.0304	0.1708	0.6560	
14	0.0020	1.0000	0.0278	0.3800	0.6560	99.52
15	0.0017	0.0000	0.0255	0.3538	0.6570	
16	0.0031	0.0000	0.0164	0.4658	0.6570	
17	0.0023	0.0000	0.0268	0.4047	0.6570	
18	0.0016	0.0000	0.0266	0.3504	0.6570	
19	0.0016	0.0000	0.0265	0.2057	0.6580	
20	0.0020	0.0000	0.0281	0.3807	0.6560	
21	0.0020	0.0000	0.0276	2.2024	0.6580	
22	0.0020	0.0000	0.0271	3.0839	0.6570	
23	0.0024	0.0000	0.0271	2.2444	0.6570	
24	0.0023	1.0000	0.0270	3.0263	0.6560	103.91
25	0.0021	0.0000	0.0270	2.7971	0.6570	
26	0.0021	0.0000	0.0410	2.8041	0.6570	
27	0.0021	0.0000	0.0517	2.9360	0.6560	
28	0.0021	0.0000	0.0404	2.7070	0.6570	
29	0.0037	0.0000	0.0517	2.5126	0.6560	
30	0.0039	0.0000	0.0172	3.0493	0.6560	
31	0.0037	0.0000	0.0308	2.9665	0.6550	
32	0.0039	1.0000	0.0283	2.8684	0.6610	120.094
33	0.0018	0.0000	0.0257	3.6488	0.6610	

34	0.0033	0.0000	0.0258	3.2960	0.6590	
35	0.0025	0.0000	0.0273	3.2805	0.6590	
36	0.0017	0.0000	0.0755	3.2832	0.6600	
37	0.0017	0.0000	0.0635	3.1907	0.6590	
38	0.0022	0.0000	0.0641	3.1768	0.6610	
39	0.0069	1.0000	0.0441	2.4176	0.6610	118.003
40	0.0043	0.0000	0.0376	2.3993	0.6600	
41	0.0043	0.0000	0.0556	2.4176	0.6600	
42	0.0038	0.0000	0.0488	2.3993	0.6590	
43	0.0038	0.0000	0.0429	2.6700	0.6600	
44	0.0040	0.0000	0.0299	2.9940	0.6630	
45	0.0040	0.0000	0.0406	3.1813	0.6630	
46	0.0069	0.0000	0.0460	2.8911	0.6630	
47	0.0035	0.0000	0.0376	2.9866	0.6620	
48	0.0076	0.0000	0.0406	3.3969	0.6620	
49	0.0076	0.0000	0.0406	2.2454	0.6630	
50	0.0053	0.0000	0.0460	2.5609	0.6640	
51	0.0076	0.0000	0.0406	2.5609	0.6660	
52	0.0076	0.0000	0.0375	2.6563	0.6660	
53	0.0076	0.0000	0.0375	2.6563	0.6640	
54	0.0076	0.0000	0.0406	2.6372	0.6640	
55	0.0053	0.0000	0.0406	2.6372	0.6630	
56	0.0053	0.0000	0.0460	2.9014	0.6630	
57	0.0076	0.0000	0.0460	3.2918	0.6620	
58	0.0076	0.0000	0.0460	3.4007	0.6630	
59	0.0076	0.0000	0.0460	3.4007	0.6630	
60	0.0076	0.0000	0.0429	2.5181	0.6640	
61	0.0076	0.0000	0.0429	2.2877	0.6640	
62	0.0076	0.0000	0.0336	2.2877	0.6650	
63	0.0069	0.0000	0.0299	2.2877	0.6660	
64	0.0069	0.0000	0.0299	2.9366	0.6630	
65	0.0048	0.0000	0.0460	2.5826	0.6640	
66	0.0035	0.0000	0.0406	2.5945	0.6630	
67	0.0035	0.0000	0.0429	2.9884	0.6640	
68	0.0076	1.0000	0.0299	2.2877	0.6650	124.87
69	0.0076	0.0000	0.0460	2.2877	0.6640	
70	0.0069	0.0000	0.0406	2.9884	0.6660	
71	0.0035	0.0000	0.0334	2.9366	0.6670	
72	0.0076	0.0000	0.0429	2.2877	0.6660	

73	0.0076	0.0000	0.0274	2.6002	0.6660	
74	0.0047	0.0000	0.0272	2.5857	0.6670	
75	0.0069	0.0000	0.0494	3.5246	0.6680	
76	0.0022	0.0000	0.0482	2.9143	0.6660	
77	0.0022	0.0000	0.0299	2.9336	0.6660	
78	0.0074	0.0000	0.0504	3.0044	0.6670	
79	0.0074	0.0000	0.0515	3.0044	0.6670	
80	0.0035	0.0000	0.0683	2.6225	0.6660	
81	0.0081	0.0000	0.0609	2.9518	0.6670	
82	0.0081	0.0000	0.0299	2.3377	0.6670	
83	0.0044	0.0000	0.0631	2.3377	0.6670	
84	0.0044	0.0000	0.0642	2.6252	0.6650	
85	0.0035	0.0000	0.0625	2.4040	0.6640	
86	0.0078	0.0000	0.0578	3.4227	0.6630	
87	0.0047	0.0000	0.0563	3.3441	0.6630	
88	0.0078	0.0000	0.0692	2.3990	0.6630	
89	0.0096	1.0000	0.0610	2.3990	0.6630	127.012
90	0.0096	0.0000	0.0611	2.8521	0.6640	
91	0.0047	0.0000	0.0660	2.5734	0.6640	
92	0.0113	0.0000	0.0648	2.5734	0.6620	
93	0.0113	0.0000	0.0728	2.7975	0.6620	
94	0.0086	0.0000	0.0719	2.7975	0.6610	
95	0.0086	0.0000	0.0779	2.9021	0.6590	
96	0.0101	0.0000	0.0778	2.4612	0.6600	
97	0.0101	0.0000	0.0715	2.8234	0.6610	
98	0.0085	0.0000	0.0651	2.4612	0.6590	
99	0.0085	0.0000	0.0715	2.3276	0.6620	
100	0.0080	0.0000	0.0691	2.3276	0.6620	
101	0.0080	0.0000	0.0770	2.8234	0.6610	
102	0.0080	0.0000	0.0770	2.4074	0.6630	
103	0.0080	1.0000	0.0817	2.4074	0.6670	126.975
104	0.0119	0.0000	0.0804	2.6310	0.6680	
105	0.0119	0.0000	0.0690	2.6310	0.6660	
106	0.0094	0.0000	0.0692	2.3280	0.6670	
107	0.0094	0.0000	0.0894	2.3280	0.6670	
108	0.0048	0.0000	0.0890	2.4347	0.6670	
109	0.0048	0.0000	0.0642	2.4347	0.6670	
110	0.0111	0.0000	0.0692	2.7213	0.6670	
111	0.0111	0.0000	0.0343	2.7213	0.6680	

112	0.0051	0.0000	0.0343	2.7213	0.6670	
113	0.0051	0.0000	0.0343	2.7213	0.6680	
114	0.0049	0.0000	0.0343	2.3980	0.6670	
115	0.0049	0.0000	0.0945	2.3980	0.6670	
116	0.0049	0.0000	0.0945	2.5963	0.6670	
117	0.0049	1.0000	0.0827	2.5963	0.6670	127.991
118	0.0131	0.0000	0.0877	3.0588	0.6670	
119	0.0131	0.0000	0.0579	3.0588	0.6670	
120	0.0093	0.0000	0.0573	2.3280	0.6680	
121	0.0093	0.0000	0.0611	2.3280	0.6690	
122	0.0125	0.0000	0.0592	3.1265	0.6700	
123	0.0125	1.0000	0.0730	3.1265	0.6720	129.57
124	0.0133	0.0000	0.0731	3.5358	0.6730	
125	0.0133	0.0000	0.0950	3.2093	0.6720	
126	0.0131	0.0000	0.0900	3.2231	0.6720	
127	0.0131	0.0000	0.0690	3.2292	0.6720	
128	0.0121	0.0000	0.0692	2.3798	0.6720	
129	0.0121	0.0000	0.0877	2.3798	0.6720	
130	0.0053	0.0000	0.0877	2.8091	0.6720	
131	0.0053	0.0000	0.0771	2.8091	0.6730	
132	0.0102	0.0000	0.0751	2.4905	0.6740	
133	0.0102	0.0000	0.0465	2.4905	0.6750	
134	0.0140	1.0000	0.0715	2.6560	0.6760	133.75
135	0.0140	0.0000	0.0897	2.6560	0.6770	
136	0.0080	0.0000	0.0902	2.5238	0.6770	
137	0.0080	0.0000	0.0947	2.5238	0.6770	
138	0.0144	0.0000	0.0925	2.6797	0.6770	
139	0.0144	0.0000	0.0950	2.6797	0.6780	
140	0.0154	0.0000	0.0950	3.3609	0.6780	
141	0.0154	0.0000	0.0827	3.3609	0.6760	
142	0.0142	0.0000	0.0877	2.9629	0.6750	
143	0.0142	0.0000	0.0900	2.9629	0.6760	
144	0.0102	0.0000	0.0950	2.6835	0.6750	
145	0.0102	0.0000	0.0628	2.6835	0.6740	
146	0.0157	0.0000	0.0877	3.2203	0.6750	
147	0.0157	0.0000	0.0877	3.2203	0.6760	
148	0.0187	0.0000	0.0900	2.3935	0.6740	
149	0.0113	0.0000	0.0950	2.3935	0.6730	
150	0.0113	0.0000	0.0950	2.5440	0.6730	

151	0.0133	0.0000	0.0950	2.5440	0.6730	
152	0.0133	0.0000	0.0900	2.7138	0.6740	
153	0.0168	1.0000	0.0950	2.7138	0.6750	133.93
154	0.0168	0.0000	0.0950	3.2155	0.6750	
155	0.0172	1.0000	0.0950	3.2155	0.6750	135.675
156	0.0172	0.0000	0.0587	2.8146	0.6750	
157	0.0156	0.0000	0.0582	2.8146	0.6750	
158	0.0156	0.0000	0.0613	2.5713	0.6760	
159	0.0188	0.0000	0.0615	3.3599	0.6780	
160	0.0188	0.0000	0.0595	3.3599	0.6770	
161	0.0204	0.0000	0.0747	3.2074	0.6770	
162	0.0204	0.0000	0.0734	3.2074	0.6780	
163	0.0239	0.0000	0.0584	2.9561	0.6780	
164	0.0198	0.0000	0.0950	2.9561	0.6770	
165	0.0198	0.0000	0.0950	3.0082	0.6770	
166	0.0239	0.0000	0.0617	3.0082	0.6760	
167	0.0184	0.0000	0.0617	3.2201	0.6750	
168	0.0184	0.0000	0.0774	3.2201	0.6760	
169	0.0256	0.0000	0.0776	2.9561	0.6770	
170	0.0256	0.0000	0.0900	2.9561	0.6760	
171	0.0216	0.0000	0.0921	3.0422	0.6780	
172	0.0216	1.0000	0.0950	3.0422	0.6780	139.57
173	0.0218	0.0000	0.0755	2.5426	0.6770	
174	0.0218	0.0000	0.0950	2.8977	0.6770	

Table 2.
Claim amounts statistics, paths and jump forecasts.

Author details

Leonard Mushunje*, Chiedza Elvina Mashiri, Edina Chandiwana and Maxwell Mashasha
Department of Applied Mathematics and Statistics, Midlands State University, Gweru, Zimbabwe

*Address all correspondence to: leonsmushunje@gmail.com

References

[1] Merton, R. C. (1975). Optimum consumption and portfolio rules in a continuous-time model. In: Stochastic Optimization Models in Finance (pp. 621–661). Academic Press.

[2] Merton, R. C. (1976). Option pricing when underlying stock returns are discontinuous. Journal of Financial Economics, 3, 125–144. DOI:10.1016/0304-405X(76) 90022-2.

[3] Bollerslev, T., Todorov, V., and Li, S. Z. (2013). Jump tails, extreme dependencies, and the distribution of stock returns. Journal of Econometrics, 172(2), 307–324. DOI:10.1016/j.jeconom.2012.08.014

[4] Cremers, M., Halling, M., and Weinbaum, D. (2015). Aggregate jump and volatility risk in the cross-section of stock returns. The Journal of Finance, 70(2), 577–614. DOI:10.1111/jofi.12220

[5] Jarrow, R. A., and Rosenfeld, E. R. (1984). Jump risks and the intertemporal capital asset pricing model. Journal of Business, 57(3), 337-351.

[6] Jorion, P. (1989). Asset allocation with hedged and unhedged foreign stocks and bonds. Journal of Portfolio Management, 15(4), 49.

[7] IFE (2019) Institute and Faculty of Actuaries (IFoA) (2019). Actuarial Statistics Notes with revision questions.

[8] Merton, R. C. (1973). An intertemporal capital asset pricing model. Econometrica, 41(5), 83. DOI: 10.2307/1913811, Pierre (2017)

[9] Ross, S. A. (1976). Options and efficiency. The Quarterly Journal of Economics, 90(1), 75-89.

[10] Fama, E. F. (1973). A note on the market model and the two-parameter model. Journal of Finance, 48, 1181–1185. DOI:10.1111/j.1540-6261.1973.tb01449.x

[11] Black, F and Scholes, M. (1973). The pricing of options and corporate liabilities. Journal of Political Economy Vol. 81, No. 3 (May–Jun., 1973), pp. 637-654 (18 pages) Published By: The University of Chicago Press. https://www.jstor.org/stable/1831029

Chapter 6

Fuzzy Perceptron Learning for Non-Linearly Separable Patterns

Raja Kishor Duggirala

Abstract

Perceptron learning has its wide applications in identifying interesting patterns in the large data repositories. While iterating through their learning process perceptrons update the weights, which are associated with the input data objects or data vectors. Though perceptrons exhibit their robustness in learning about interesting patterns, they perform well in identifying the linearly separable patterns only. In the real world, however, we can find overlapping patterns, where objects may associate with multiple patterns. In such situations, a clear-cut identification of patterns is not possible in a linearly separable manner. On the other hand, fuzzy-based learning has its wide applications in identifying non-linearly separable patterns. The present work attempts to experiment with the algorithms for fuzzy perceptron learning, where perceptron learning and fuzzy-based learning techniques are implemented in an interfusion manner.

Keywords: perceptron learning, fuzzy-based learning, fuzzy C-means, interfusion, weighted distances, pattern recognition, sum of squared errors, clustering fitness

1. Introduction

A learning system could be thought as a collection of methods that are brought together in order to create an environment to facilitate different learning processes. The learning systems will provide various types of learning resources and descriptions of procedures for obtaining quality results [1]. The learning systems find their applications in the areas like, image recognition, speech recognition, traffic prediction, e-mail spam and malware filtering, automatic language translation, medical diagnosis, etc. [2].

As the data increases in large volumes in the digital repositories, it has become essential to look for alternative approaches to yield better results in extracting interesting patterns from the repositories. Intelligent learning systems are gaining attention from a wide range of researchers in the recent years in extracting patterns from the data repositories. The learning systems have three kinds of approaches. They are supervised, unsupervised, and semi-supervised learning approaches [3].

The concept of perceptron learning plays a critical role in pattern recognition, which has become a challenging problem in the data science research. In the recent years, perceptron learning algorithms are exhibiting their robust performance in identifying interesting patterns from large data repositories when compared to the traditional supervised learning approaches [4]. A perceptron can be thought as a computational prototype of a neuron. As a supervised learning approach, perceptron learning is used for linear classification of patterns. This learning approach uses the already available labelled data to classify the future data by predicting the class labels.

In the literature, it is studied that many researchers experimented with perceptron learning for identifying interesting patterns from the data. A novel autonomous perceptron model (APM) was proposed to address the issues of complexity of traditional perceptron architectures [4]. APM is a nonlinear supervised learning model, which has the architecture using the computational power of the quantum bits (qubits). The researchers [5], using biophysical perceptron (BP), tried to simulate the pyramidal cells in the brain with a wide variety of active dendritic channels. The BP, here, explores the ability of real neurons with extended non-linear dendritic trees to effectively perform the classification task in identifying interesting patterns from the data. Many researchers have experimented with perceptron learning in a wide variety of ways. However, the perceptron learning suffers several limitations. It works well for linearly separable patterns. Though some researchers experimented for identifying non-linearly separable patterns, the perceptron learning produced best results for binary separation of patterns only [5]. Also that perceptron learning suffers poor performance in case of overlapping patterns, that is, when patterns are not having sharp boundaries.

Fuzzy-based learning, on the other hand, is found to show its ability in performing well for overlapping patterns [6]. As a fuzzy-based learning approach, fuzzy C-means (FCM) is widely used by researchers for pattern recognition. A weighted local fuzzy regression model showed a better efficiency than the least squares regression for non-linear and high-dimensional pattern recognition of transport system in China [7]. The new kernelized fuzzy C-means clustering algorithm [8] uses a kernel-induced distance function as a similarity measure showed improved performance in identifying the patterns when compared to the conventional fuzzy C-means technique. In many research findings, it is observed that the fuzzy-based learning approach was used in a wide variety of ways to achieve better results in extracting non-linear and overlapping patterns.

The present work attempts to experiment with fuzzy perceptron learning, which implements the perceptron learning and fuzzy-based learning techniques in an interfusion manner. In the research literature, we can find a good amount of work related to the combination of fuzzy logic with perceptron learning. The fuzzy neural network (FNN) was proposed for pattern classification, which uses supervised fuzzy clustering and pruning algorithm to determine the precise number of clusters with proper centroids representing the patterns to be recognised [9]. In the fuzzy neural integrated networks [10], the researchers attempted to integrate the concept of fuzzy sets and neural networks to deal with pattern recognition problems. In an enhanced algorithm for fuzzy lattice reasoning (FLR) classifier, a new nonlinear positive valuation function was defined to produce better results for pattern classification [11]. Along with these, however, many other research experiments of fuzzy perceptron learning are supervised learning approaches only. Therefore, the present work focuses on experimenting with effective implementation of some techniques involved in the perceptron and fuzzy-based learning systems for unsupervised learning to identify interesting patterns in large datasets. As part of the present work, five algorithms are developed, two of which are related to perceptron learning, one is the standard fuzzy C-means (FCM) algorithm. The remaining two algorithms are proposed by the present work, which implement the perceptron learning and fuzzy-based learning in an interfusion manner using weights and weighted distances respectively. All the algorithms are implemented using three benchmark datasets. The CPU time, clustering fitness (CF), and sum of squared errors (SSE) are taken into consideration for performance evaluation of the algorithms.

2. Perceptron learning

Nowadays, the perceptron learning model can be thought as a more general computational model in identifying interesting patterns in a dataset. It takes an input, aggregates it along with the weights and produces the result. A perceptron is used to learn patterns and relationships in data. Patterns help us knowing about the interesting features around which objects may be grouped in a given population of data.

A perceptron may be configured for a specific application, such as pattern recognition and data classification through some learning process [12]. Perceptrons are information processing devices, which are built from interconnected elementary processing units. These units are called neurons. The perceptrons are robust in exhibiting their ability in distributed representation and computation, learning, generalisation, adaptivity, inherent contextual information processing, and fault tolerance [13].

The perceptron learning uses an iterative weight adjustment for the enhanced retrieval of patterns from a dataset. The iterative process converges to the weights, which produce the patterns that represent the different groups of data objects in the dataset uniquely. While operating for learning on patterns, the perceptrons use weights in connection to every input vector. A weight represents the information used by the perceptron to solve a problem [14].

The perceptron with multiple neurons is shown in **Figure 1**.

In **Figure 1**, $X_1, X_2, \dots, X_n$ are the n input vectors and $Y_1, Y_2, \dots, Y_m$ are the m neurons. The input vector X_1 is connected to neurons $Y_1, Y_2, \dots, Y_m$ with weights W_{11}, $W_{12}, \dots, W_{1m}$, respectively, the input vector X_2 is connected to the neurons with weights $W_{21}, W_{22}, \dots, W_{2m}$, respectively so on and the input vector X_n is connected to the neurons with the weights $W_{n1}, W_{n2}, \dots, W_{nm}$, respectively. The weights of all input vectors for all neurons will be formulated as the weight matrix as shown below.

$$\begin{bmatrix} W_{11} & W_{12} & \dots & \dots & W_{1m} \\ W_{21} & W_{22} & \dots & \dots & W_{2m} \\ \dots & \dots & \dots & \dots & \dots \\ \dots & \dots & \dots & \dots & \dots \\ \dots & \dots & \dots & \dots & \dots \\ W_{n1} & W_{n2} & \dots & \dots & W_{nm} \end{bmatrix}$$

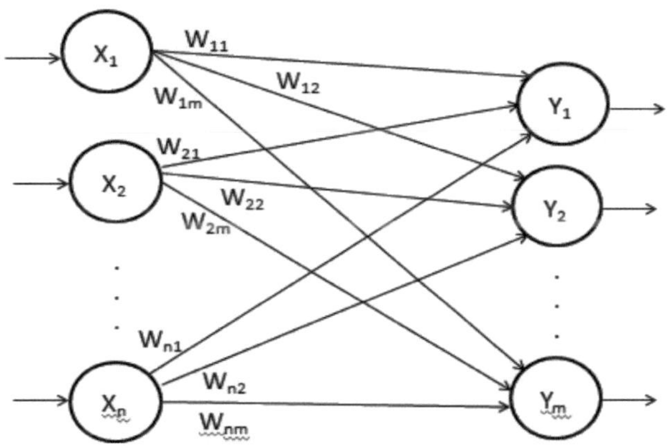

Figure 1.
A perceptron with a multiple neurons.

Though the perceptron learning exhibits its robustness in identifying the patterns in the data repositories, it works well for linearly separable patterns, that is, the patterns with sharp boundaries only. However, in the real time world, we may find overlapping patterns, that is, non-linearity in pattern associativity, where data objects may associate with multiple patterns. In such situations, the perceptron learning approach may suffer in identifying the patterns clearly. On the other hand, fuzzy-based learning has its wide applications in identifying patterns in the overlapping scenario. In the present work, two algorithms are implemented for perceptron learning. They are discussed in the following sub-sessions.

2.1 Perceptron learning using weights (PLW)

This algorithm implements the perceptron learning using weights [12]. With each input data vector, a weight is associated corresponding to each pattern. To generate the initial weights, one iteration of K-means algorithm is performed. The results of K-means iteration are used to compute the weight matrix. This weight matrix will be repeatedly updated in the subsequent iterations. For each input data vector weights are computed corresponding to every pattern. The input data vector is associated with the pattern corresponding to which the weight is maximum. This process is repeated for every iteration. The algorithm terminates when there is no change in the association of data vectors to the patterns. The algorithm for perceptron learning using weights is given below.

2.1.1 Algorithm PLW

Step 1: Determine the number of patterns, k, to be recognised from the dataset.
Step 2: Select k points randomly from the dataset and set them as cluster seeds to correspond the patterns to be recognised.
Step 3: Perform one iteration of K-means algorithm.
Step 4: Using the results of K-means iteration, compute cluster wise initial weights.
Step 5: Repeat steps 6–8 until the stopping condition.
Step 6: Generate weight matrix, where each element W_{ij} is computed as:

$$W_{ij}(i+1) = W_{ij}(i) + lr * \left(\|X_i\| - W_{ij}(i)\right) \tag{1}$$

Here, $W_{ij}(i+1)$ is the weight of ith data point X_i for jth cluster for the iteration $(i+1)$, $W_{ij}(i)$ is the weight of ith data point for jth cluster for the iteration i, $\|X_i\|$ is the norm of data point X_i, and lr is the learning rate. The lr may assume a value ranging between 0 and 1. To avoid possible biasedness in the computations, lr is assumed to be 0.5.
Step 7: Assign points to clusters using weights.
Step 8: Update cluster means, that is, refine patterns.
[End of step 5 loop]
Step 9: [End of algorithm]

2.2 Perceptron learning using weighted distances (PLWD)

This algorithm implements the perceptron learning using weighted distances [15]. With each input data vector, a weighted distance is associated corresponding to each pattern. To generate the initial weighted distances, one iteration of K-means algorithm is performed. Using the results of K-means the weight matrix is computed. This weight matrix is used to compute the weighted distances for each input data vector. The data vector is associated with the pattern corresponding to which the weighted distance is minimum. This weight matrix will be repeatedly updated in the subsequent iterations to compute the new weighted distances. This process repeats for every iteration. The algorithm terminates when there is no change in the association of data vectors to the patterns. The algorithm for perceptron learning using weighted distances is given below.

2.2.1 Algorithm PLWD

Step 1: Determine the number of patterns, k, to be recognised from the dataset.
Step 2: Select k points randomly from the dataset and set them as cluster seeds μ_j (j = 1, 2, ..., m) to correspond the patterns to be recognised.
Step 3: Perform one iteration of K-means algorithm.
Step 4: Using the results of K-means iteration, compute cluster wise initial weights.
Step 5: Repeat steps 6–10 until the stopping condition.
Step 6: Generate weight matrix W using Eq. (1).
Step 7: For each data point X_i, compute the Euclidean distance $d(X_i, \mu_j)$ as follows:

$$d\left(X_i, \mu_j\right) = \sqrt{\sum_{l=1}^{d} \left(x_{il} - \mu_{jl}\right)^2} \tag{2}$$

Here, X_i is the ith data point, μ_j is the mean vector of the cluster j.
Step 8: For each data point compute the weighted distances as follows:

$$Wd_j = W_{ij}(i+1) \cdot d\left(X_i, \mu_j\right) \tag{3}$$

Step 9: Assign points to clusters using weights.
Step 10: Update cluster means, that is, refine patterns.
[End of step 5 loop]
Step 11: [End of algorithm]

Though the perceptron learning algorithms are experimented widely by many researchers, they exhibit their robustness in identifying linearly separable patterns only.

3. Fuzzy-based learning

Fuzzy-based learning is used to handle the concept of partial truth, where the truth value may range between completely true and completely false [16]. It is an approach that allows for multiple possible truth values to be processed through the same data object. In fuzzy-based learning, the data objects are assumed being associated with multiple patterns. For each data object, the degree of association is measured in membership. This membership value may range between 0 and 1 (1 being high similarity and 0 being no similarity with the pattern).

Fuzzy-based learning techniques focus on modelling uncertain and vague information that is found in the real world situations. These techniques deal with the patterns whose boundaries cannot be defined sharply [17, 18]. By fuzzy-based learning, one can know if data objects fully or partially associate with the patterns that are under consideration based on their memberships of association [19]. Among the techniques of fuzzy-based learning, fuzzy C-means (FCM) is the most well-known one as it has the advantage of robustness for obscure information about the patterns [20, 21]. FCM is widely studied and applied in geological shape analysis [22], medical diagnosis [23], automatic target recognition [24], meteorological data [20], pattern recognition, image analysis, image segmentation and image clustering [25–27], agricultural engineering, astronomy, chemistry [28], detection of polluted sites [29], etc. The following section presents a brief discussion of FCM algorithm.

3.1 Fuzzy C-means (FCM)

The fuzzy C-means (FCM) is a technique that uses degree of membership for natural interpretation of patterns recognised [30]. The FCM associates the data vectors among *k* patterns. Each data vector may associate with each pattern with a membership degree. The membership of a data vector towards a pattern can range between 0 and 1.

The FCM algorithm is given below [31]. Here, U is the $k \times N$ membership matrix. While computing the cluster means and updating the membership matrix at each iteration, the FCM uses the fuzzifier factor, m. For most cases, m ranging between 1.5 and 3.0 gives good results [32]. In the present work, in all the experiments, m is set to 1.5.

3.1.1 Algorithm FCM

Step 1: Determine the number of patterns, k, to be recognised from the dataset.

Step 2: Select k points randomly from the dataset and set them as cluster seeds μ_j ($j = 1, 2, \ldots, m$) to correspond the patterns to be recognised.

Step 3: Perform one iteration of K-means algorithm. Set $t = 0$.

Step 4: Using the results of K-means iteration, compute membership matrix $U^{(0)}_{k \times N}$.

Step 5: Repeat steps 6–9 until the stopping condition.

Step 6: [Refine patterns] Update the mean of jth cluster μ_j as follows:

$$\mu_j = \frac{\sum_{i=1}^{N} (u_{ij})^m X_i}{\sum_{i=1}^{N} (u_{ij})^m} \tag{4}$$

Here, u_{ij} is the membership degree of the data point X_i w.r.t. jth pattern and m is the fuzzifier factor.

Step 7: Compute the new membership matrix using:

$$u_{ij}{}^{t+1} = \left[\sum_{l=1}^{k} \left(\frac{\left\| X_i - \mu_j{}^t \right\|^2}{\left\| X_i - \mu_l{}^t \right\|^2} \right)^{1/m-1} \right]^{-1} \tag{5}$$

Step 9: Assign points to clusters using membership degrees. Set $t = t + 1$.

[End of step 5 loop]

Step 10: [End of algorithm]

4. Fuzzy perceptron learning

The fuzzy perceptron learning works in an interfusion manner, where the fuzzy logic is combined with perceptron learning for identifying non-linear and overlapping patterns. Much research work may be found in the literation where fuzzy perceptron learning is experimented in different applications [33, 34]. However, those experiments are confined to supervised learning only. The present work attempts to experiment with fuzzy perceptron learning for unsupervised cases. The present work proposes two algorithms, one is for fuzzy perceptron learning using weights and the other is for fuzzy perceptron learning using weighted distances.

4.1 Fuzzy perceptron learning using weights (FPLW)

This algorithm implements the perceptron learning using weights and FCM techniques in an interfusion manner. These techniques are performed in alternative iterations until the termination condition. Initially, one iteration of K-means algorithm is performed. Using the results of K-means, initial weights are computed as mentioned in the Section 2.1. Using these weights, weight matrix is generated to perform one iteration of perceptron learning algorithm to associate the input data vectors to the patterns. Using the results of perceptron learning step, membership matrix is computed to perform one iteration of FCM algorithm as mentioned in Section 3.1. The results of FCM step are used to update weight matrix to perform perceptron learning step. In this way the perceptron learning and FCM algorithms are repeated in alternative iterations until termination condition. The algorithm for fuzzy perceptron learning using weights (FPLW) is given below.

4.1.1 Algorithm FPLW

Step 1: Determine the number of patterns, k, to be recognised from the dataset.
Step 2: Select k points randomly from the dataset and set them as cluster seeds μ_j ($j = 1, 2, \ldots, m$) to correspond the patterns to be recognised.
Step 3: Perform one iteration of K-means algorithm.
Step 4: Using the results of K-means iteration, compute cluster wise initial weights.
Step-5: Update cluster means μ_j ($j = 1, 2, \ldots, m$).
Step 6: Repeat steps 7–13 until the stopping condition.
Step 7: Compute the weight matrix using Eq. (1).
Step 8: Assign points to clusters using weights.
Step 9: If there is no change in cluster assignment then go to step 14.
Step 10: Update cluster means using Eq. (4).
Step 11: Generate membership matrix $U^{(0)}_{k \times N}$ using Eq. (5).
Step 12: Assign points to clusters using membership matrix.
Step 13: If there is no change in cluster assignment then go to step 14.
[End of Step 6 loop]
Step 14: [End of Algorithm]

4.2 Fuzzy perceptron learning using weighted distances (FPLWD)

This algorithm implements the perceptron learning using weighted distances and FCM techniques in an interfusion manner. These techniques are performed in alternative iterations until the termination condition. Initially, one iteration of K-means technique is performed. Using the results of K-means, initial weights are computed as mentioned in the Section 2.2. Now, one iteration of perceptron learning algorithm is performed where the weight matrix is generated using the initial weights. Using this weight matrix, weighted distances are computed for every input vector X_i with respect to each pattern using the Eq. (3). The weighted distances are used to associate the input vectors to the patterns. Using the results of perceptron learning step, membership matrix is computed to perform one iteration of FCM algorithm as mentioned in Section 3.1. The results of FCM step are used to compute weight matrix for perceptron learning step. In this way the perceptron learning and FCM steps are repeated in alternative iterations until termination condition. The algorithm for fuzzy perceptron learning using weighted distances (FPLWD) is given below.

4.2.1 Algorithm FPLWD

Step 1: Determine the number of patterns, k, to be recognised from the dataset.
Step 2: Select k points randomly from the dataset and set them as cluster seeds μ_j (j = 1, 2, … , m) to correspond the patterns to be recognised.
Step 3: Perform one iteration of K-means algorithm.
Step 4: Using the results of K-means iteration, compute cluster wise initial weights.
Step-5: Update cluster means μ_j (j = 1, 2, … , m).
Step 6: Repeat steps 7–15 until the stopping condition.
Step 7: Compute the weight matrix using Eq. (1).
Step 8: For each data point compute the Euclidean distance using Eq. (2).
Step 9: For each data point compute weighted distances using Eq. (3).
Step 10: Assign points to clusters using weighted distances.
Step 11: If there is no change in cluster assignment then go to step 16.
Step 12: Update cluster means using Eq. (4).
Step 13: Generate membership matrix $U^{(0)}_{k \times N}$ using Eq. (5).
Step 14: Assign points to clusters using membership matrix.
Step 15: If there is no change in cluster assignment then go to step 16.
[End of Step 6 loop]
Step 16: [End of Algorithm]

5. Performance evaluation

For performance evaluation of algorithms, CPU time in seconds, sum of squared errors [35] and clustering fitness (CF) [36] are taken into consideration and are calculated for all the algorithms.

5.1 Sum of squared errors

The objective of pattern learning is to minimise the intra-cluster sum of squared errors (SSE). The lesser the SSE, the better the goodness of fit is. The SSE for the results of each algorithm is computed using Eq. (6).

$$SSE = \sum_{j=1}^{k} \sum_{X_i \in C_j} \left(X_i - \mu_j\right)^2 \tag{6}$$

Here, X_i is the ith data point in the dataset, μ_j (j = 1, … , k) is the mean of the cluster C_j, and k is the number of patterns to be recognised.

5.2 Cluster fitness

While achieving high intra-cluster similarity, it is also important to achieve well separation of patterns.

So, it is also important to consider inter-cluster similarity while evaluating the performance of the algorithms. For this, the present work, computes the clustering fitness (CF) as a performance criterion, which requires the calculation of both intra-cluster similarity and inter-cluster similarity. The computation of CF also requires the experiential knowledge, λ. The computation of CF results in higher value when the inter-cluster similarity is low and results in lower value for when the inter-cluster similarity is high. Also that to make the computation of CF unbiased, the value of λ is taken as 0.5 [36].

5.2.1 Intra-cluster similarity for the cluster C_j

It can be quantified via a function of the reciprocals of intra-cluster radii within each of the resulting clusters. The intra-cluster similarity of a cluster C_j $(1 = j = k)$, denoted as $S_{tra}(C_j)$ [36], is defined by:

$$S_{tra}(C_j) = \frac{1+n}{1+\sum_1^n dist(I_l, Centroid)} \tag{7}$$

Here, n is the number of items in cluster C_j, I_j $(1 = j = n)$ is the jth item in cluster C_j, and dist(I_j, *Centroid*) calculates the distance between I_j and the centroid of C_j, which is the intra-cluster radius of C_j. To smooth the value of $S_{tra}(C_j)$ and allow for possible singleton clusters, 1 is added to the denominator and numerator.

5.2.2 Intra-cluster similarity for one clustering result C

It is denoted as $S_{tra}(C)$ [36]. It is defined by:

$$S_{tra}(C) = \frac{\sum_1^k S_{tra}(C_j)}{k} \tag{8}$$

Here, k is the number of resulting clusters in C and $S_{tra}(C_j)$ is the intra-cluster similarity for the cluster C_j.

5.2.3 Inter-cluster similarity

It can be quantified via a function of the reciprocals of inter-cluster radii of the clustering centroids. The inter-cluster similarity for one of the possible clustering results C, denoted as $S_{ter}(C)$ [36] is defined by:

$$S_{ter}(C) = \frac{1+k}{1+\sum_1^k dist\left(Centroid_j, Centroid^2\right)} \tag{9}$$

Here, k is the number of resulting clusters in C, $1 = j = k$, $Centroid_j$ is the centroid of the jth cluster in C, $Centroid^2$ is the centroid of all centroids of clusters in C. We compute inter-cluster radius of $Centroid_j$ by calculating dist($Centroid_j$, $Centroid^2$), which is distance between $Centroid_j$, and $Centroid^2$. To smooth the value of $S_{ter}(C)$ and allow for possible all-inclusive clustering result, 1 is added to the denominator and the numerator.

5.2.4 Clustering fitness

The clustering fitness for one of the possible clustering results C, denoted as CF [36], is defined by:

$$CF = \lambda \text{ x } S_{tra}(C) + \frac{1-\lambda}{S_{ter}(C)} \tag{10}$$

Here, λ $(0 < \lambda < 1)$ is an experiential weight, $S_{tra}(C)$ is the intra-cluster similarity for the clustering result C and $S_{ter}(C)$ is the inter-cluster similarity for the clustering result C.

6. Experiments and results

Experimental work has been carried out on the system with Intel(R) Core(TM) i3-5005 U CPU@2.00GHz processor speed, 4GB RAM, Windows 7 OS (64-bit) and using JDK1.7.0_45. Separate modules are written for each of the above discussed methods to observe the CPU time for clustering any dataset by keeping the cluster seeds same for all methods. I/O operations are eliminated and the CPU time observed is strictly for clustering of the data.

Along with the proposed algorithms FPLW and FPLWD for fuzzy perceptron learning, experiments are also conducted with the algorithms PLW, PLWD and FCM for performance comparison. All the algorithms are executed using the benchmark datasets with varying number of patterns to be recognised. In the present work, Magic Gamma, Letter Recognition and Intrusion datasets are used from UCI ML data repository [37]. All the developed algorithms, PLW, PLWD, FCM, FPLW and FPLWD, are executed using these datasets for varying number of patterns to be recognised (k = 10, 11, 12, 13, 14, 15).

All the algorithms operate in an iterative manner and terminate when a stopping condition is met. The stopping condition is when there is no change in the pattern associativity of the data vectors. The termination condition is the same for all the algorithms.

Details of the datasets are available in **Table 1**.

S. No.	Dataset	No. of points	No. of dimensions
1	Magic Gamma data	19,020	10
2	Letter Recognition data	20,000	16
3	Intrusion data	4,94,019	35

Table 1.
Details of datasets.

6.1 Observations with Magic Gamma dataset

The results of all algorithms, using Magic Gamma dataset, with respect to CPU time in seconds, clustering fitness and sum of squared errors are shown in **Figures 2–4**, respectively.

6.2 Observations with Letter Recognition dataset

The results of all algorithms, using Letter Recognition dataset, with respect to CPU time in seconds, clustering fitness and sum of squared errors are shown in **Figures 5–7**, respectively.

6.3 Observations with Intrusion dataset

The results of all algorithms, using Intrusion dataset, with respect to CPU time in seconds, clustering fitness and sum of squared errors are shown in **Figures 8–10**, respectively.

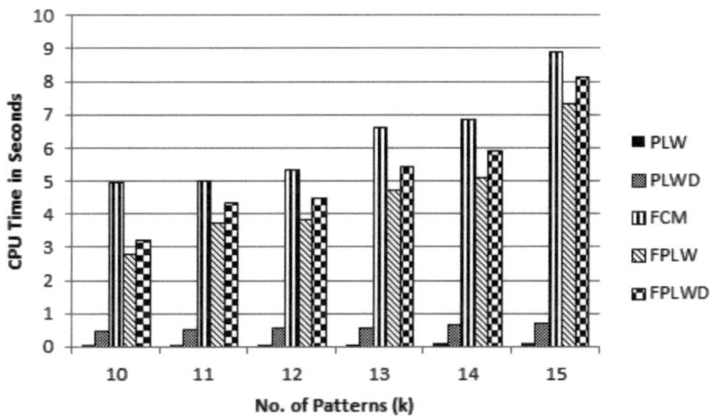

Figure 2.
CPU time of each clustering method (Magic Gamma dataset).

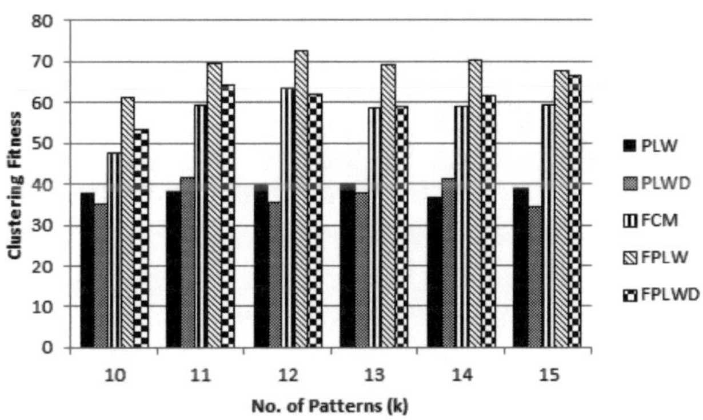

Figure 3.
Clustering fitness of each clustering method (Magic Gamma dataset).

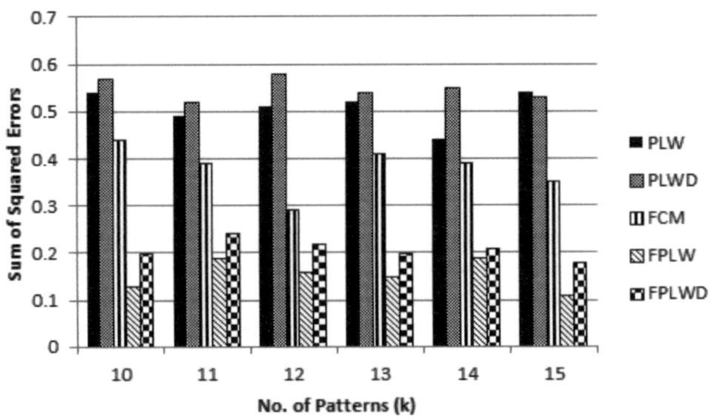

Figure 4.
SSE of each clustering method (Magic Gamma dataset).

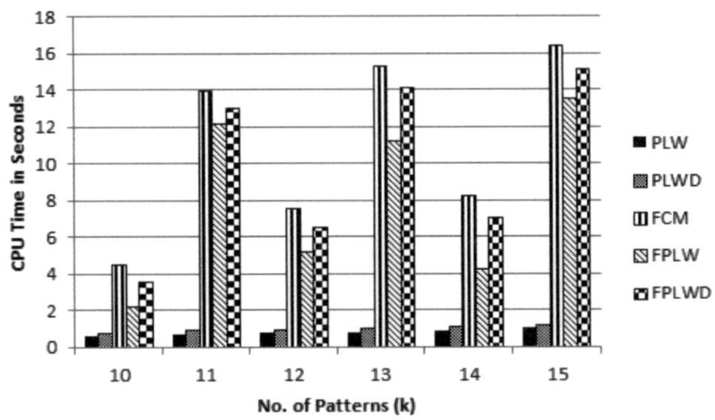

Figure 5.
CPU time of each clustering method (Letter Recognition dataset).

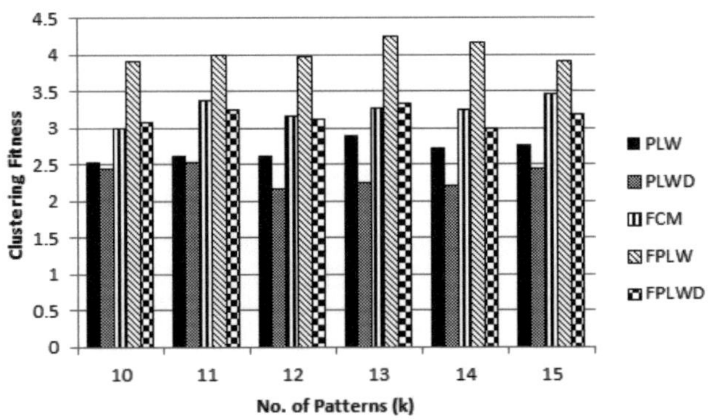

Figure 6.
Clustering fitness of each clustering method (Letter Recognition dataset).

In all the experiments, it is observed that the algorithm FPLW, which implements the perceptron learning using weights and the FCM techniques in an interfusion manner, is showing consistently better performance in terms of clustering fitness (CF) and SSE than the other algorithms.

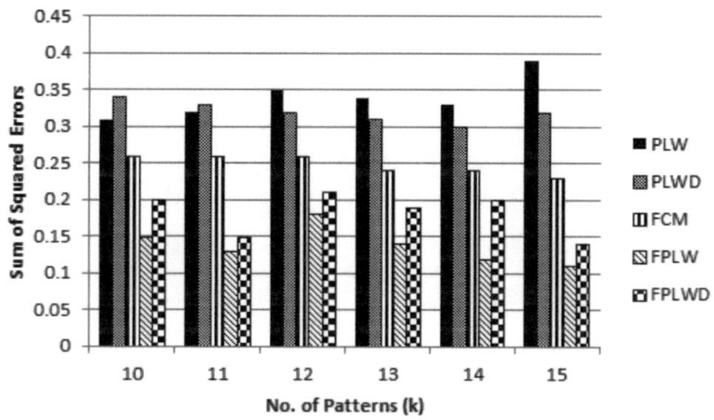

Figure 7.
SSE of each clustering method (Letter Recognition dataset).

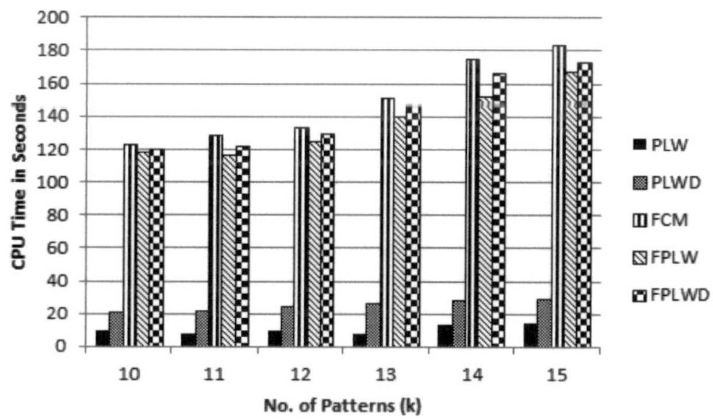

Figure 8.
CPU time of each clustering method (Intrusion dataset).

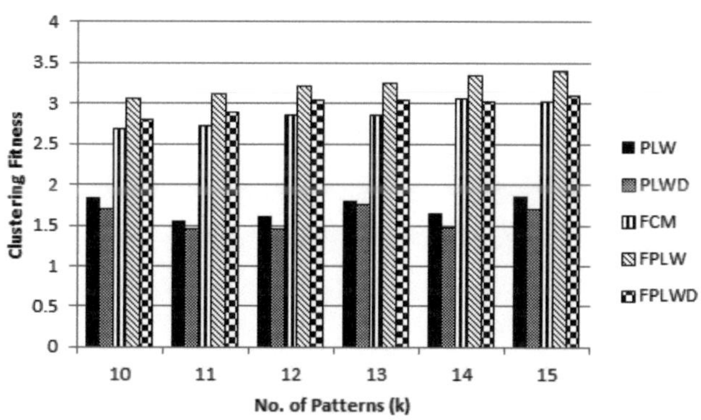

Figure 9.
Clustering fitness of each clustering method (Intrusion dataset).

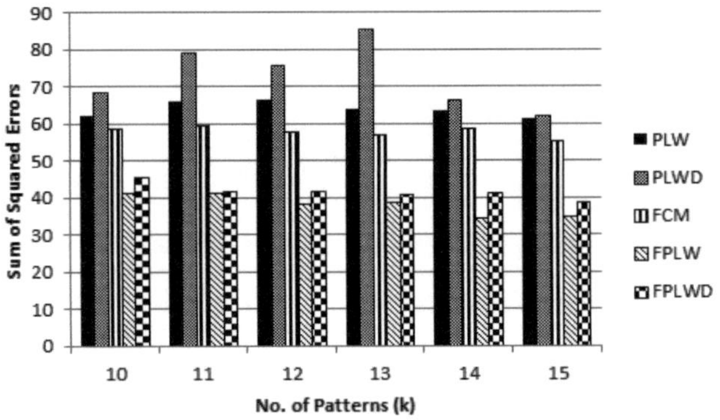

Figure 10.
SSE of each clustering method (Intrusion dataset).

7. Conclusion

The present experiment mainly focuses on the study fuzzy perceptron learning for recognising non-linear patterns in the datasets. Many researchers contributed greatly towards fuzzy perceptron learning. However, their experiments are confined to supervised learning only. So, the present work experimented with the fuzzy perceptron learning approaches for unsupervised learning. The work proposes two new algorithms, that is, FPLW and FPLWD. These algorithms are implemented using three benchmark datasets. Along with these algorithms, the algorithms for standard FCM and perceptron learning using weights and weighted distances are also implemented for performance comparison. For all the algorithms the CPU time in seconds, clustering fitness (CF) and sum of squared errors (SSE) are taken into consideration for performance evaluation. All the developed algorithms are experimented with varying number of patterns (k) to be recognised.

In all the experiments, it is observed that the proposed algorithm for fuzzy perceptron learning using weights (FPLW) is consistently showing better performance with respect to clustering fitness and SSE. Of course, the algorithm FPLW is taking a little more time for its execution than the other algorithms. However, it could be negligible, as the main concern is for clearly recognising the non-linear patterns in the datasets.

Author details

Raja Kishor Duggirala
Department of Computer Science Engineering, Dr. Lankapalli Bullayya College of Engineering, Visakhapatnam, Andhra Pradesh, India

*Address all correspondence to: rajakishor@gmail.com

References

[1] Satapathy SK et al. EEG Brain Signal Classification for Epileptic Seizure Disorder Detection, ScienceDirect, 2019, ISBN 978-0-12-817426-5, DOI: https://doi.org/10.1016/C2018-0-01888-5

[2] Benton WC, Hat R, Raleigh. Machine learning systems and intelligent applications. IEEE Software. 2020;**37**: 43-49. DOI: 10.1109/MS.2020.2985224

[3] Krendzelak M. Machine Learning and Its Applications in e-Learning Systems. Stary Smokovec, Slovakia: IEEE, 2014 IEEE 12th IEEE International Conference on Emerging eLearning Technologies and Applications (ICETA); 2014. pp. 267-269. DOI:10.1109/ICETA.2014.7107596

[4] Sagheer A, Zidan M, Abdelsamea MM. A novel autonomous perceptron model for pattern classification applications. Entropy. 2019;**21**(8):763. DOI: 10.3390/e21080763

[5] Moldwin T, Segev I. Perceptron learning and classification in a modeled cortical pyramidal cell. Frontiers in Computational Neuroscience. 2020. DOI: 10.3389/fncom.2020.00033. https://www.frontiersin.org/articles/10.3389/fncom.2020.00033/full

[6] Leoni Sharmila S, Dharuman C, Venkatesan P. A fuzzy based classification—An experimental analysis, International Journal of Innovative Technology and Exploring Engineering. 2019;**8**(10):4634-4638

[7] Tan Y, Chen S. Pattern recognition based on weighted fuzzy C-means clustering. In: 6th International Congress on Image and Signal Processing (CISP). 2013. pp. 1061-1065. DOI: 10.1109/*CISP*.2013.6745213

[8] Das S, Baruah H. A new kernelized fuzzy C-means clustering algorithm with enhanced performance. Semantic Scholar; 2014. Corpus ID: 212563388

[9] Kulkarni A, Kulkarni N. Fuzzy neural networks for pattern recognition. Procedia Computer Science. 2020;**167**:2606-2616. DOI: 10.1016/j.procs.2020.03.321

[10] Baraldi A, Blonda P, Petrosino A. Fuzzy neural networks for pattern recognition. In: Marinaro M, Tagliaferri R, editors. Neural Nets WIRN VIETRI-97. Perspectives in Neural Computing. London: Springer; 1998. DOI: 10.1007/978-1-4471-1520-5_2

[11] Jamshidi Khezeli YJ, Nezamabadi-pour H. Fuzzy lattice reasoning for pattern classification using a new positive valuation function. Advances in Fuzzy Systems. 2012;**2012**:206121. DOI: 10.1155/2012/206121

[12] Sivanandam SN, Sumathi S, Deepa SN. Introduction to Neural Networks Using Matlab 6.0. India: Tata McGraw Hill; 2008

[13] Nadal J-P, Parga N. Information processing by a perceptron in an unsupervised learning task. Network: Computation in Neural Systems. 1993; **4**(3):295-312. DOI: 10.1088/0954-898X_4_3_004

[14] Haykin S. Neural Networks: A Comprehensive Foundation. 2nd ed. New Delhi, India: Pearson Education; 2007

[15] Nalaie K, Ghiasi-Shirazi K, Akbarzadeh-T M. Efficient implementation of a generalized convolutional neural networks based on weighted Euclidean distance. In: 2017 7th International Conference on Computer and Knowledge Engineering (ICCKE). 2017. pp. 211-216. DOI: 10.1109/*ICCKE*.2017.8167877

[16] Novák V, Perfilieva I, Močkoř J. Mathematical Principles of Fuzzy Logic. Dordrecht: Kluwer Academic; 1999

[17] Zadeh LA. Fuzzy sets. Information and Control. 1965;**8**(3):338-353

[18] Lemiare J. Fuzzy insurance. ASTIN Bulletin. 1990;**20**(1):33-55

[19] Das S. Pattern recognition using fuzzy C-means technique. International Journal of Energy Information and Communications. 2013;**4**(1):1-14

[20] Lu Y, Ma T, Yin C, Xie X, Tian W, Zhong SM. Implementation of the fuzzy C-means clustering algorithm in meteorological data. International Journal of Database Theory and Application. 2013;**6**(6):1-18

[21] Kaltri K, Mahjoub M. Image segmentation by Gaussian mixture models and modified FCM algorithm. The International Arab Journal of Information Technology. 2014;**11**(1):11-18

[22] Bezdek JC, Trivedi M, Ehrlich R, Full WE. Fuzzy clustering: A new approach for geostatistical analysis. International Journal of System Measurement and Decision. 1981;**1**:13-23

[23] Bezdek JC. Feature selection for binary data-medical diagnosis with fuzzy sets. In: Proc. Nat. Comput. Conf. AFIPS Press; 1972. pp. 1057-1068

[24] Cannon RL, Dave JV, Bezdek JC. Efficient implementation of fuzzy C-means clustering algorithm. IEEE Transactions on Pattern Analysis and Machine Intelligence. 1986;**8**(2):248-255

[25] Cannon RL, Jacobs C. Multispectral pixel classification with fuzzy objective functions. Technical Report CAR-TR-51. College Park: Center for Automation Research, University of Maryland; 1984

[26] Gong M, Liang Y, Ma W, Ma J. Fuzzy C-means clustering with local information and kernel metric for image segmentation. IEEE Transactions on Image Processing. 2013;**22**(2):573-584

[27] Krinidis S, Krinidis M, Chatzis V. Fast and robust fuzzy active contours. IEEE Transactions on Image Processing. 2010;**19**(5):1328-1337

[28] Yong Y, Chongxun Z, Pan L. A novel fuzzy C-means clustering algorithm for image thresholding. Measurement Science Review. 2004;**4**(1):11-19

[29] Hanesch M, Scholger R, Dekkers MJ. The application of fuzzy C-means cluster analysis and non-linear mapping to a soil data set for the detection of polluted sites. Physics and Chemistry of the Earth, Part A. 2001;**26**(11–12):885-891

[30] Ghosh S, Dubey SK. Comparative analysis of K-means and fuzzy C-means algorithms. International Journal of Advanced Computer Science and Applications. 2013;**4**(4):35-39

[31] Klir GJ, Yuan B. Fuzzy Sets and Fuzzy Logic: Theory and Applications. India: Prentice Hall of India Private Limited; 2005

[32] Bezdek JC, Ehrlich R, Full W. FCM: The fuzzy C-means clustering algorithm. Computers and Geosciences. 1984;**10** (2–3):191-203

[33] Auephanwiriyakul S, Dhompongsa S. An investigation of a linguistic perceptron in a nonlinear decision boundary problem. In: 2006 IEEE International Conference on Fuzzy Systems. 2006. pp. 1240-1246

[34] Yang J, Wu W, Shao Z. A new training algorithm for a fuzzy perceptron and its convergence. In: Advances in Neural Networks—ISNN 2005, Second

International Symposium on Neural Networks, Chongqing, China; May 30–June 1 2005

[35] Han J, Kamber M. Data Mining Concepts and Techniques. 2nd ed. San Francisco, CA: Morgan Kaufmann Publishers, An Imprint of Elsevier; 2007

[36] Han X, Zhao T. Auto-K dynamic clustering algorithm. Journal of Animal and Veterinary Advances. 2005;**4**(5): 535-539

[37] Lichman M. UCI Machine Learning Repository. 2013. Available from: http://archive.ics.uci.edu/ml

Chapter 7

Semantic Map: Bringing Together Groups and Discourses

Theodore Chadjipadelis and Georgia Panagiotidou

Abstract

This chapter presents a multivariate analysis method which is developed in two steps using a combination of Hierarchical cluster analysis (HCA) and Factorial Correspondence Analysis (AFC). To explain and describe the steps of the method, we use an application example on a survey dataset from young students in Thessaloniki trying to investigate their behavioral profiles in terms of political characteristics and how these may be affected about their attendance to a civic education course offered by the Political Science department in the Aristotle University of Thessaloniki. The method is explained step by step on this example serving as a manual of its application to the researcher. HCA assigns subjects into cluster membership variables and in the next stage, these new variables are jointly analyzed with AFC. Correspondence analysis manages to extract the dimensions of the phenomenon in the study, explaining the inner antithesis between the categories but also giving the opportunity to visualize the information in a two-dimensional space, a semantic map, making interpretation more comprehensive. HCA is then applied again to the AFC's coordinates of the categories constructing profiles of subjects, assigning them to the categories of the variables.

Keywords: hierarchical cluster analysis, correspondence analysis, political analysis, multivariate methods, data analysis

1. Introduction

This chapter presents a multivariate analysis method, using a combination of Hierarchical Cluster Analysis (HCA) [1] and Factorial Correspondence Analysis (AFC) in two steps [2]. The method provides the advantage of jointly handling multiple variables with many levels. The approach exploits HCA in reducing many variables into fewer ones that represent the individuals within them and then with Correspondence analysis it manages to reduce the information even further and express it upon dimensions.

These dimensions not only organize the information within the data to be explained more thoroughly but also visualizes the inner relationships among categories of the variables. By analyzing the antagonism of the clusters on different sets of dimensions, as we can also have a three-dimensional or more system of axes [3], we can understand further the behavior of the variables and their categories, as well as the associations among them.

Clustering in the final step of the coordinates of the categories on the dimensions we link the initial clusters with the categories, creating a semantic map [4] that can visualize the phenomenon in a Cartesian field or a three-dimensional space [3]. In this chapter, we present the application of the method in a specific case, which works only as an example.

The sample consists of students in Thessaloniki, Greece measuring specifically their political attitudes and their views on democracy, on moral values and the way they are informed in general about politics. In the example that is developed through the chapter we describe the application of the method and the interpretation of the results step by step.

2. Methodology

Our data analysis is based on Hierarchical Cluster Analysis (HCA) and Factorial Correspondence Analysis (AFC) in two steps [5]. The dataset is analyzed using advanced multivariate methods (Hierarchical Cluster analysis, Factorial correspondence Analysis (Analyse factorielle des correspondences AFC) [2]. Using this mixed-method approach, enables the detection of profiles of similar behavior, the association of each profile to the distinct categories that compose it and the detection of the dimension which describes the dynamics of the phenomenon, enabling the visualization of these dynamics in its final output.

In the first step, HCA assigns subjects into distinct groups according to their response patterns [2]. The main output of HCA is a group or cluster membership variable, which reflects the partitioning of the subjects into groups. Furthermore, for each group, the contribution of each question (variable) to the group formation is investigated [2], to reveal a typology of behavioral patterns. To determine the number of clusters, we use the empirical criterion of the change in the ratio of between-cluster inertia to total inertia, when moving from a partition with r clusters to a partition with r-1 clusters [6]. The metric used is chi-square. Analysis was conducted with the software M.A.D. (Methodes de l' Analyse des Donnees) [7]. In the second step, the group membership variable, obtained from the first step, is jointly analyzed with the existing variables via Multiple Correspondence Analysis on the so-called Burt table [8]. At this stage, correspondence extracts the dimensions that constitute the overall phenomenon, explaining the inner inertia between all subjects. To determine the number of factors, the empirical criterion of Benzecri was used. According to the empirical criterion of Benzecri [2], two specific sub-criteria should be fulfilled.

$COR > 200$ and CTR value $>1000/(n + 1)$.

where n = total number of categories.

We proceed by applying again HCA for the coordinates of the categories on the dimensions. Bringing these two analyses steps together, we can construct a semantic map that can visualize the behavioral structure of the variables and the subjects, creating behavioral patterns and abstract discourses [4].

3. An application example in political analysis

To demonstrate the method of HCA and MCA in two steps, an example was selected to be described in the following sections. This example refers to the analysis of data collected during a survey in Thessaloniki, Greece in the period 2019–2020. The topic of the survey is to collect data about the political characteristics of young students who participated in a civic education course offered by the Department of Political Sciences in the Aristotle University of Thessaloniki. The sample consists of 1618 participants, allocated into four groups:

Group 1: random university students within the campus of the university who were not part of the civic education course.

Group 2: university students who attended the course in-classroom.

Group 3: university students who attended the course through e-learning, due to covid-19 restrictions and measures.

Group 4: high-school students who attended the course.

The tool of the survey was a questionnaire, structured in three sections: 1) demographics, 2) political behavior, 3) information means, views on democracy and moral context.

The objective of the research is to investigate the students' levels of political knowledge, political interest, preferable way of political mobilization and distinguish the different profiles among the four groups of participants. The variables of the research -associated with each one of the questions- correspond to: a) political interest, c) political knowledge, b) political mobilization, c) their self-positioning on the ideological left–right axis, d) sources of information on politics e) structure of the "political" and f) "moral" self [9, 10].

More specifically, the respondents are asked directly for their level of political interest (ordinal scale) and the way they prefer to mobilize themselves on political issues which may arise (nominal scale). The variable of political knowledge (ordinal scale) is composed through the answers of the respondents on basic questions about politics, many correct answers produce a high score of political knowledge. Next, the respondents are asked to position themselves on a scale of 0 to 10 resembling the left–right ideological axis.

In the last section of the questionnaire, the questions on information sources, democratic and moral self are found. Regarding the preferable source of information, the respondents are asked to choose the two sources they use more often to get informed about politics. Moving on to the variable of "democratic self" [10], the respondent finds a set of 12 pictures, which conceptualize different versions of democracy. They are asked to choose three of them that symbolize in the best way how they understand democracy. Same wise, in the next question they are asked again to choose 3 pictures from a new set of 12 pictures, representing attitudes and views on life and moral values in general. These two sets of pictures construct symbolic representations of democratic institutions and of their personal moral compass (**Table 1**) [9].

3.1 First step of the analysis: clustering subjects into distinct groups

In this step of the analysis, we select the three variables of the last section, these are the sources of information (E13), the understanding of democracy (E14) and the moral values (E15). For these variables, we have a dataset comprising of 0–1 values, where 0 equals to a not selected picture or source and 1 to a selected one. For each one of these three sets of variables, we apply HCA, aiming to summarize the information. HCA's output is the dendrogram in **Figure 1** visualizing the clusters created in each step.

Initially, we cluster the variables to see patterns of categories. In the example below, we cluster the pictures for democracy, getting 5 clusters (38, 40, 41, 46 and 44). As seen in **Figure 2**, cluster 38 is created by the selection of pictures 3, 10 and 11, cluster 40 consists of selecting picture 1 etc.

Code	Variable	Categories				
group	**group**	**1: random students**	**2: students in-class**	**3: students e-learning**	**4: high school students**	
lr_c	ideology	1: left	2: left-left	3: left	4: left-right	5: right
PM	political mobilization (nominal)	1: I personally address the authorities	2: I participate with others in collective mobilizations	3: I take action through Social Media	4: I let the authorities to do their job	5: I do not know / I do not answer
PI	political interest (ordinal)	1: very much	2: quite	3: a little	4: not at all	
PK	political knowledge (ordinal)	1: low	2: moderate	3: high		
E13	political info source (categorical, binary 0–1)	1: TV-Radio	2: Online newspapers-Internet	3: Social media	4: Family-relatives	5: Friends 6: Newspapers
E14	perception of democracy (categorical, binary 0–1)	12 pictures which visualize concepts for how they perceive democracy				
E15	personal values (categorical, binary 0–1)	12 pictures which visualize concepts of moral values				

Table 1.
Coding and categories of the variables used in the analysis.

Processing the same HCA analysis, to cluster the variables for each one of the three selected variables, we get 5 clusters for E14, 5 clusters for E15 and 4 clusters for E13, as shown in the **Table 2**.

We proceed by clustering now the subjects. Instead of having 12 binary variables to represent the democratic self, we produce clusters of similar choices and assign each one of the respondents to the clusters he is closer to according to this profile of answers. HCA again produces a dendrogram with the steps of the clustering process (**Figure 3**).

In the example shown in **Figure 4** we see how the answers on the 12 pictures on democratic self are transformed into one clustering variable (gr_dem), assigning each respondent into one of the clusters of HCA. Following the same method, a separate application of HCA for information sources and for the moral self we get the clustering variables (gr_inf) and (gr_val).

After we have completed a separate HCA, to classify the subjects (respondents) for each one of the selected variables (E14, E15 and E13) we get 8 clusters of respondents for E14 (renamed to gr_dem), 9 clusters for E15 (renamed to gr_val) and 8 clusters for E13 (renamed to gr_inf). **Table 3** shows a summary of the clusters of subjects for each

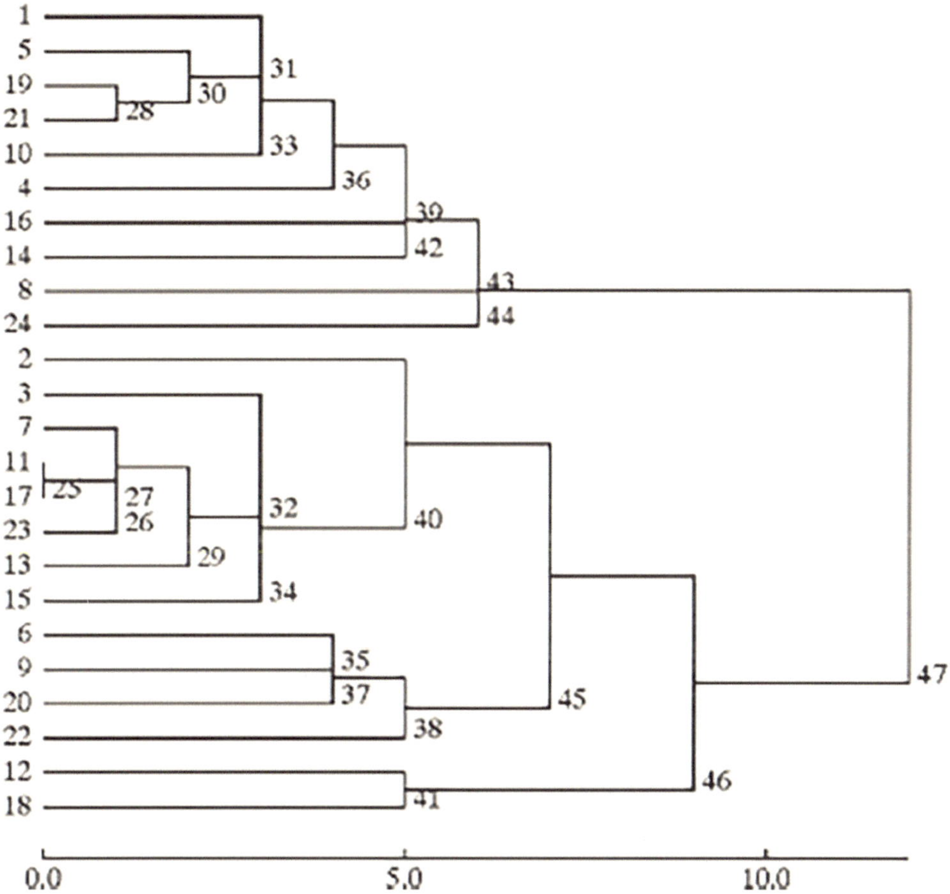

Figure 1.
Dendrogram (HCA) indicating the clusters for E14 variable.

Cluster	A(I)	B(I)	Βάρος										
25	11	17	0,1542	E1460	E1490								
26	25	23	0,2261	E1460	E1490	E14120							
27	7	26	0,29546	E1440	E1460	E1490	E14120						
28	19	21	0,12652	E14100	E14110								
29	27	13	0,36042	E1440	E1460	E1490	E14120	E1470					
30	5	28	0,18704	E1430	E14100	E14110							
31	1	30	0,24394	E1410	E1430	E14100	E14110						
32	3	29	0,411	E1420	E1440	E1460	E1490	E14120	E1470				
33	31	10	0,28227	E1410	E1430	E14100	E14110	E1451					
34	32	15	0,46164	E1420	E1440	E1460	E1490	E14120	E1470	E1480			
35	6	9	0,06781	E1431	E1450								
36	33	4	0,31501	E1410	E1430	E14100	E14110	E1451	E1421				
37	35	20	0,09113	E1431	E1450	E14101							
38	37	22	0,10794	E1431	E1450	E14101	E14111						
39	36	16	0,3477	E1410	E1430	E14100	E14110	E1451	E1421	E1481			
40	2	34	0,48807	E1411	E1420	E1440	E1460	E1490	E14120	E1470	E1480		
41	12	18	0,01246	E1461	E1491								
42	39	14	0,36606	E1410	E1430	E14100	E14110	E1451	E1421	E1481	E1471		
43	42	8	0,38002	E1410	E1430	E14100	E14110	E1451	E1421	E1481	E1471	E1441	
44	43	24	0,39145	E1410	E1430	E14100	E14110	E1451	E1421	E1481	E1471	E1441	E14121
45	40	38	0,59601	E1411	E1420	E1440	E1460	E1490	E14120	E1470	E1480	E1431	E1450
46	45	41	0,60846	E1411	E1420	E1440	E1460	E1490	E14120	E1470	E1480	E1431	E1450
47	44	46	0,99992	E1410	E1430	E14100	E14110	E1451	E1421	E1481	E1471	E1441	E14121

Figure 2.
Classification process of the 12 pictures-variables of E14 (from E141 to E1412).

Clusters for democracy	**38**	**40**	**41**	**44**	**46**
pictures selected	E1431	E1411	E1461	E1451	
	E14101		E1491	E1421	
	E14111			E1481	
				E1471	
				E1441	
				E14121	
Clusters for values	37	38	39	40	41
pictures selected	E1531	E1591	E1511	E1571	E1541
	E1581		E1521	E15121	E1551
				E1561	
clusters for information	23	24	25	26	
pictures selected	e1311	e1361	e1321	e1381	
	e1331	e1351			
		e1341			

Table 2.
The clusters for each one of the variables (E4, E15 and E13) and the selected pictures they are linked to.

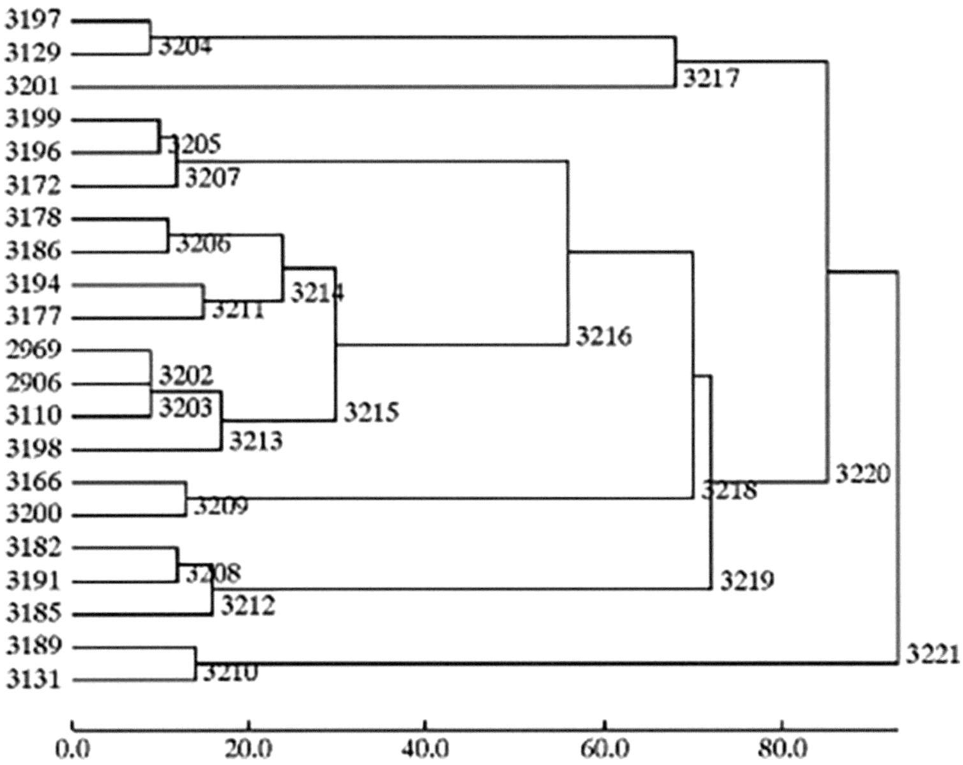

Figure 3.
Dendrogram (HCA) indicating the clusters of subjects for E14 variable.

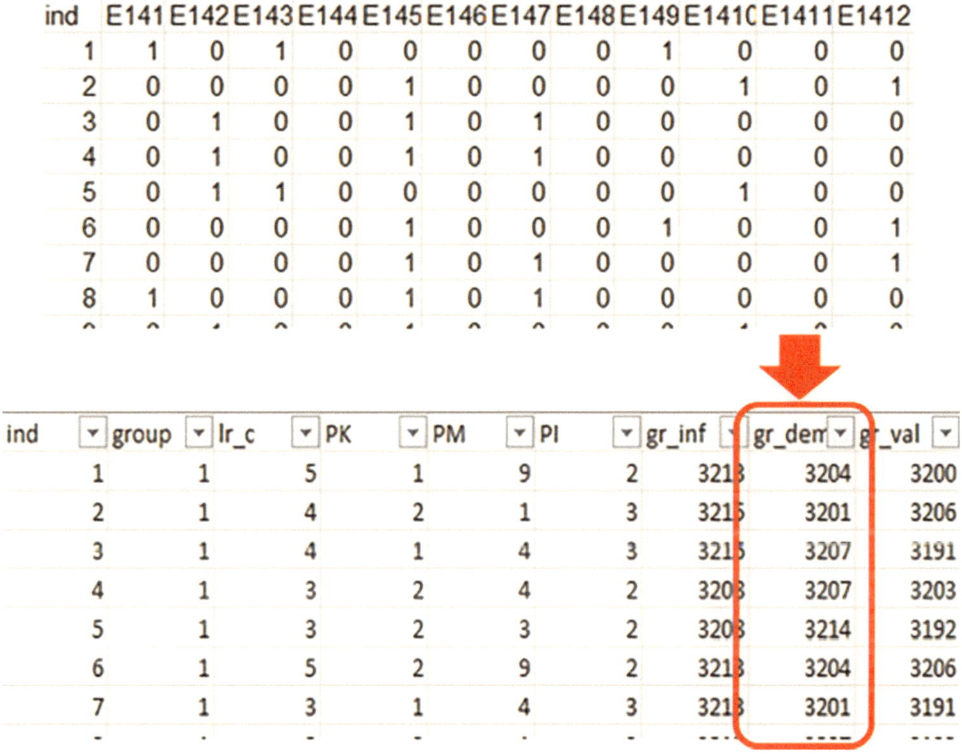

ind	E141	E142	E143	E144	E145	E146	E147	E148	E149	E1410	E1411	E1412
1	1	0	1	0	0	0	0	0	1	0	0	0
2	0	0	0	0	1	0	0	0	0	1	0	1
3	0	1	0	0	1	0	1	0	0	0	0	0
4	0	1	0	0	1	0	1	0	0	0	0	0
5	0	1	1	0	0	0	0	0	0	1	0	0
6	0	0	0	0	1	0	0	0	1	0	0	1
7	0	0	0	0	1	0	1	0	0	0	0	1
8	1	0	0	0	1	0	1	0	0	0	0	0

ind	group	lr_c	PK	PM	PI	gr_inf	gr_dem	gr_val
1	1	5	1	9	2	3213	3204	3200
2	1	4	2	1	3	3215	3201	3206
3	1	4	1	4	3	3215	3207	3191
4	1	3	2	4	2	3203	3207	3203
5	1	3	2	3	2	3203	3214	3192
6	1	5	2	9	2	3213	3204	3206
7	1	3	1	4	3	3213	3201	3191

Figure 4.
Transforming the dataset by replacing the binary E141-E1412 with the cluster membership variable gr_dem.

gr_dem	freq%	gr_val	freq%	gr_inf	freq%
3201	12%	3187	4%	3136	11%
3204	7%	3191	7%	3198	4%
3207	14%	3192	10%	3206	12%
3209	14%	3200	9%	3208	15%
3210	6%	3202	13%	3211	15%
3212	15%	3203	15%	3213	13%
3213	14%	3204	12%	3215	16%
3214	18%	3206	13%	3216	13%
		3207	16%		

Table 3.
Cluster membership variables and their categories for E14, E15 and E13.

one of the three variables we get the following table including the clusters and their relative frequency.

We investigate further the profile of each cluster for the variable E14. Each cluster is associated with selecting a set of pictures. As shown in **Table 4** cluster 3201 consists of the respondents who are more likely to select picture number 12, which corresponds to the symbolic representation for religion (**Table 5**). Cluster 3204 relates to

E14/gr_dem	3201	3204	3207	3209	3210	3212	3213	3214
E1411						40,451	27,1001	
E1421			21,534				18,6383	27,784
E1431						57,626		20,2902
E1441		41,865		82,1471				
E1451		16,437	11,9273				17,5035	
E1461					154,0449			
E1471			67,2476					
E1481							11,8125	11,1089
E1491		122,6539			22,5595			
E14101					30,127	10,2896		34,607
E14111					13,3978	71,9056		
E14121	93,9969	7021						

Table 4.
Weight of selecting each picture to the creation of the clusters for E14.

Democracy		3201	3204	3207	3209	3210	3212	3213	3214
		dem_1	dem_2	dem_3	dem_4	dem_5	dem_6	dem_7	dem_8
Movement	E1411						X	X	
Ancient Greece	E1421			X				X	X
Direct	E1431						X		X
e-Democracy	E1441		X		X				
Representative	E1451		X	X				X	
Riot	E1461					X			
Deliberation	E1471			X					
Volunteerism	E1481							X	X
Clientelism	E1491		X			X			
Rebellion	E14101					X	X		X
Protest	E14111					X	X		
Religion	E14121	X	X						
%Count		11.9%	7.4%	14.0%	13.6%	6.0%	15.0%	14.2%	18.1%

Table 5.
Summarizing the content of each cluster and renaming the clusters for E14.

selecting pictures 4,5,9 and 12 (e-democracy, representative, clientelism and religion). The sets of pictures connected to the clusters, depict the different profiles of the respondents according to the way they comprehend democracy.

Similarly, for variable E15, we describe the profiles of the cluster of the respondents regarding the pictures they are more likely to select. In **Table 6** we see that cluster 3187 is connected to the pictures 1, 2, 4 and 11 which correspond to riot, anonymous, army and protest, a representation of expressivist moral values (**Table 7**). In contrast, we see

E15/gr_val	3187	3191	3192	3200	3202	3203	3204	3206	3207
E1511	188,512								
E1521	20,9268	118,983		54,775					
E1531				12,0584				58,0211	
E1541	18,172			121,4029					
E1551				79,232			74,1092		
E1561			73,3846						
E1571			10,2587			48,654			15,5155
E1581								18,882	15,8182
E1591							44,774	82,153	23,8393
E15101						74,8128			
E15111	22,4576	52,603			80,7176				
E15121			23,373						19,8778

Table 6.
Weight of selecting each picture to the creation of the clusters for E15.

cluster 3207 having a completely naturalist moral values as it is connected to pictures 7, 8, 9, 12 (mountain, family, intimacy and concert).

Once more, we investigate the content of each cluster for the variable E13, regarding sources of information. Cluster 3136 includes those respondents who answer 1 and 3 (**Table 8**) which translates into preferring to get informed about politics by TV-radio and family (**Table 9**).

3.2 Second step: joint analysis of the cluster membership variables

In the second step of the analysis, we jointly analyze the initial variables together with the new cluster membership variables gr_dem, gr_var and gr_inf. We repeat the steps as in the early stages of the analysis applying HCA which produced the following clusters for the subjects, as w result 8 clusters of respondents are detected (**Table 10**).

These clusters relate to the categories of the variables creating a behavioral profile for each one of the clusters of the respondents, in which they have been assigned accordingly. In **Table 11** the profiles of the clusters are given in full detail, e.g., cluster 3155 consists of respondents who belong to group 4, are men [sex1], they characterize themselves as center-left [lr_c2], have moderate political knowledge [PK2], they choose to mobilize by personally addressing the authorities, take action through social media and/or let the authorities to do their job [PM1, PM3 and/or PM4],have a little political interest [PI3]. Furthermore, respondents in this cluster belong also in cluster 3136, 3208 and 3216 on how they get informed on politics, they belong to clusters 3207,3209, 3213 and 3214 regarding their views on democracy, and finally they belong in cluster 3192 regarding their set of moral values.

In the same way, we continue to examine each one of the clusters of the respondents to understand their behavioral profile, considering the total number of the variables used in our analysis.

In the next step, with the application of correspondence analysis, we extract the dimensions of the analysis and a set of coordinates for each one of the dimensions for each one of the variable categories (**Table 12**).

Values	Picture		3187	3191	3192	3200	3202	3203	3204	3206	3207
			val_1	val_2	val_3	val_4	val_5	val_6	val_7	val_8	val_9
Expressivist	Riot	E1511	X								
Expressivist	Anonymous	E1521	X	X		X					
Christian	Christ	E1531				X				X	
Army	Army	E1541	X			X					
Naturalist	Money	E1551				X			X		
Moon exploration	Astronaut	E1561			X						
Spirituality	Mountain	E1571			X			X			P
Naturalist	Family	E1581								X	P
Naturalist	Intimacy	E1591							X	X	P
Spirituality	Meditation	E15101						X			
Expressivist	Protest	E15111	X	X			X				
Naturalist	Concert	E15121			X						P
%Count			4.2%	7.2%	9.8%	8.6%	13.4%	15.4%	12.4%	13.5%	15.6%

Table 7.
Summarizing the content of each cluster and renaming the clusters for E15.

E13/gr_inf	3136	3198	3206	3208	3211	3213	3215	3216
e1311	29,2252			21,2416	29,2252			
e1321						97,3186		
e1331	52,5758							40,4936
e1341							81,3803	
e1351			38,546		26,4659			78,426
e1361			34,7882	36,0722				
e1381		181,9963						

Table 8.
Weight of selecting each source of information to the creation of the clusters for E13.

Info Source		3136	3198	3206	3208	3211	3213	3215	3216
		inf_1	inf_2	inf_3	inf_4	inf_5	inf_6	inf_7	inf_8
TV-Radio	e1311	X			X	X			
Newspapers	e1321						X		
Family	e1331	X							X
Friends	e1341							X	
Social Media	e1351			X		X			X
internet	e1361			X	X				
No information	e1381		X						
%Count		11.0%	4.5%	12.3%	15.1%	14.6%	13.0%	15.9%	13.5%

Table 9.
Summarizing the content of each cluster and renaming the clusters for E13.

Cluster	Freq%
3155	5%
3170	6%
3174	6%
3177	8%
3185	38%
3187	11%
3192	17%
3194	8%

Table 10.
Clustering for the subjects using all the variables together with the new cluster membership variables, produced in the first step.

An extra but final step of HCA is applied this time on the coordinates of the categories classifying them into groups (**Figure 5**).

The analysis highlights the existence of 10 distinct discourses of behavior (**Table 13**):

	3155	3170	3174	3177	3187	3194	3185	3192
group1			11,8463			25,592	82,319	26,596
group2		148,5301						
group3				125,4921				
group4	10,9198				14,5687	34,229		55,459
sex1	76,276					29,353		14,1511
sex2		23,639	71,799	75,205	6207		20,234	
lr_c1		30,833	38,7565	90,106		95,594		
lr_c2	93,414	83,845	10,519					
lr_c3					48,308		62,839	
lr_c4		5919			85,217			
lr_c5					2274	79,067		14,2434
PK0						79,321	53,899	
PK1			98,862		33,421	16,697	25,422	27,604
PK2	15,5399	76,055			3605			26,049
PK3		18,3509			50,077			19,997
PK9				125,4921				
PM1	34,843	78,408					45,907	
PM2		73,697	34,2217	30,939		58,304		22,388
PM3	92,603				21,341			
PM4	38,361			21,751	69,315			29,301
PM9						18,0682	25,506	
PI1		21,7371	99,778	23,305				57,641
PI2		86,385	47,958	87,025				16,576
PI3	81,295				68,028		61,683	
PI4						31,3818	1344	
gr_inf3136	35,692		17,137		40,6787			40,057
gr_inf3198						95,479		
gr_inf3206		17,9617	72,514	18,0647				
gr_inf3208	12,1097	13,512		72,484			3135	
gr_inf3211					60,759		16,943	35,994
gr_inf3213		10,4056	20,2326					17,819
gr_inf3215							31,174	
gr_inf3216	68,826						68,826	25,316
gr_dem3201					45,4001			
gr_dem3204								36,4704
gr_dem3207	23,595	16,3147		38,131			36,193	
gr_dem3209	54,348			60,254			45,489	
gr_dem3210						36,1486		17,7273

DOI: http://dx.doi.org/10.5772/intechopen.103818

	3155	**3170**	**3174**	**3177**	**3187**	**3194**	**3185**	**3192**
gr_dem3212			48,5952					
gr_dem3213	36,193						76,896	
gr_dem3214	14,6412	61,185	54,366	58,628				
gr_val3187						92,596		
gr_val3191								38,8079
gr_val3192	114,2029			20,676				
gr_val3200								35,4314
gr_val3202		22,3301	82,6908	36,965				
gr_val3203		25,968		61,137			12,4995	
gr_val3204							46,966	58,266
gr_val3206				20,116	39,9724			
gr_val3207					5277		11,5083	

Table 11.
Association between the clusters produced in the second step and the categories of the analysis.

categories	**x**	**y**
group1	−135	18
group2	415	−276
group3	1192	865
group4	−179	−221
sex1	−22	−160
sex2	15	112
lr_c1	726	−466
lr_c2	78	−13
lr_c3	−235	156
lr_c4	−75	141
lr_c5	−38	−308
PK0	−208	79
PK1	−165	−27
PK2	−94	−86
PK3	−18	−236
PK9	1192	865
PM1	−45	139
PM2	494	−432
PM3	−12	1
PM4	−186	193
PM9	−211	−41
PI1	712	−262

categories	x	y
PI2	113	9
PI3	−201	85
PI4	−414	−17
inf_1	−270	−24
inf_2	−518	−167
inf_3	381	181
inf_4	160	112
inf_5	−219	25
inf_6	283	−206
inf_7	15	19
inf_8	−198	−70
dem_1	−217	31
dem_2	−269	−105
dem_3	−44	215
dem_4	−1	186
dem_5	32	−703
dem_6	371	−316
dem_7	−140	140
dem_8	78	98
val_1	234	−661
val_2	86	−257
val_3	−39	194
val_4	−149	−285
val_5	550	−387
val_6	−31	309
val_7	−137	44
val_8	−160	134
val_9	−192	202

Table 12.
Coordinates for each one of the categories on two main dimensions (x,y).

Cluster	A(I)	B(I)	Weight											
51	3	16	0,01903	group3	PK9									
62	57	39	0,06357	lr_c1	PM2	gr_val3202	gr_dem3212							
72	66	29	0,06336	PM1	gr_dem3207	gr_inf3208								
84	82	43	0,12708	group1	PK1	PK0	gr_dem3209	gr_val3191						
85	83	77	0,31495	sex2	PI3	lr_c3	PK2	gr_val3207	gr_dem3213	gr_val3203	gr_inf3215	lr_c4	PM4	gr_inf3211
86	79	75	0,17293	sex1	PI2	PM3	gr_inf3216	lr_c2	gr_dem3214	gr_val3192				
87	78	65	0,05044	group2	gr_inf3206	PI1	gr_inf3213							
89	60	63	0,04217	lr_c5	gr_val3200	gr_dem3204	gr_val3204							
92	81	56	0,09893	group4	PK3	gr_inf3136	gr_dem3201	gr_val3206						
93	69	53	0,04737	PM9	PI4	gr_inf3198	gr_dem3210	gr_val3187						

Figure 5.
Clustering the variables using their coordinates on the dimension as input.

DOI: http://dx.doi.org/10.5772/intechopen.103818

10 clusters	**51**	**62**	**87**	**72**	**84**	**85**	**86**	**89**	**92**	**93**
6 clusters	51	91	91	88	90	90	88	95	95	93
4 clusters	98	98	98	94	94	94	94	95	95	93
				96	96	96	96	96	96	93
group	group3		group2		group1				group4	
Left–Right		far left				center-left/center-right	center-left	far right		
Political Interest			Very			Not very	Somewhat			Not at all
Political Knowledge	No Data				None/Little	Adequate			High	
Political Mobilization		Collective		Personal		Let others to do their job	Social Media			N/A
Gender						Female	Male			
Information Source			Social media, Internet/ Newspapers	TV-Radio, Internet		TV-Radio, social media/ Friends	Family, social media		TV-Radio, Family	No information
Democracy		Movement, Direct, Rebellion, Protest		Ancient Greece, Representative, Deliberation	e-Democracy	Movement, Ancient Greece, Representative, Volunteerism	Ancient Greece, Direct, Volunteerism, Rebellion	e-Democracy, Representative, Corruption	Religion	Riot, Corruption, Rebellion, Protest
Values		Protest			Anonymous, Protest	Spirituality, meditation/ Mountain, family, intimacy, concert	Astronaut, mountain, concert	Anonymous, Christ, money, army/ Money intimacy	Christ, family, intimacy	Riot, anonymous, army, protest

Table 13.
Summarizing the association between the categories and the clusters.

a. Clusters 51, 62, 87 which is a later step are unified in one cluster 98. This cluster reflects the profile of group 2 and 3 (university students who undertook the civic education course either in-class either online). They are characterized as far left, with high political interest, collective political mobilization, get informed by social media, internet or the newspapers. They see democracy as direct and think of it as rebellion and protest, while in their moral set of values they choose protest (expressivists).

b. Clusters 72, 84, 85, 86 which in later classification stage merge into cluster 94, including the random sample of students who were not part of the civic education course. These participants are characterized as center-left/center-right, have a moderate to low political interest, little to none political knowledge, low political mobilization (letting others do their job) or social media, they get informed by tv-radio, social media, friends and family. They view democracy as movement, representative, direct and they see a strong connection to ancient Greece. Their moral values are mainly naturalist, focusing on entertainment, family or spirituality.

c. In clusters 89, 92 which meet later in cluster 95, we find the younger high school students, who also attended the civic education course. This cluster is characterized as closer to the righter positions of the left–right axis. They demonstrate high political knowledge, they get informed by TV-radio and family and they see democracy as e-democracy, representative and connected to corruption and religion. Their moral setting is a mixture of expressivist and naturalistic values, including a set of nationalist symbolism including army, Christ, and family.

d. Cluster 93 concentrates respondents of no political interest, or information who understand democracy as rebellion or corruption and are closer to expressivist values such as riot, protest but also army.

4. Final output: the semantic map

Utilizing the coordinates of the points on the two first axes which were obtained from the correspondence analysis, we construct a system of 2 axes on which we place all these points [3]. The output resembles a simple Cartesian field where x is the first dimension (horizontal), and y is the second dimension (vertical). A third dimension can be brought into the analysis by using a three-dimensional space, visualizing the objects within a cube, or by presenting the different sets of the dimension by two.

The output is a semantic map, where all objects can be seen altogether, and their positioning on the field can be explained in terms of the object's proximity or opposition on each one of the dimensions.

In our example (**Figure 6**), we make the following observations:

The first axis is created by the opposing objects of: 1) group 1 (random students) and group 4 (high school students), followed by characteristics such as low political interest, getting informed by V-radio or friend and family, center left\center right, naturalistic values, choosing not to be mobilized or act on an individual level if needed and 2) group 2 and group 3 (university students of the civic education course) with high political

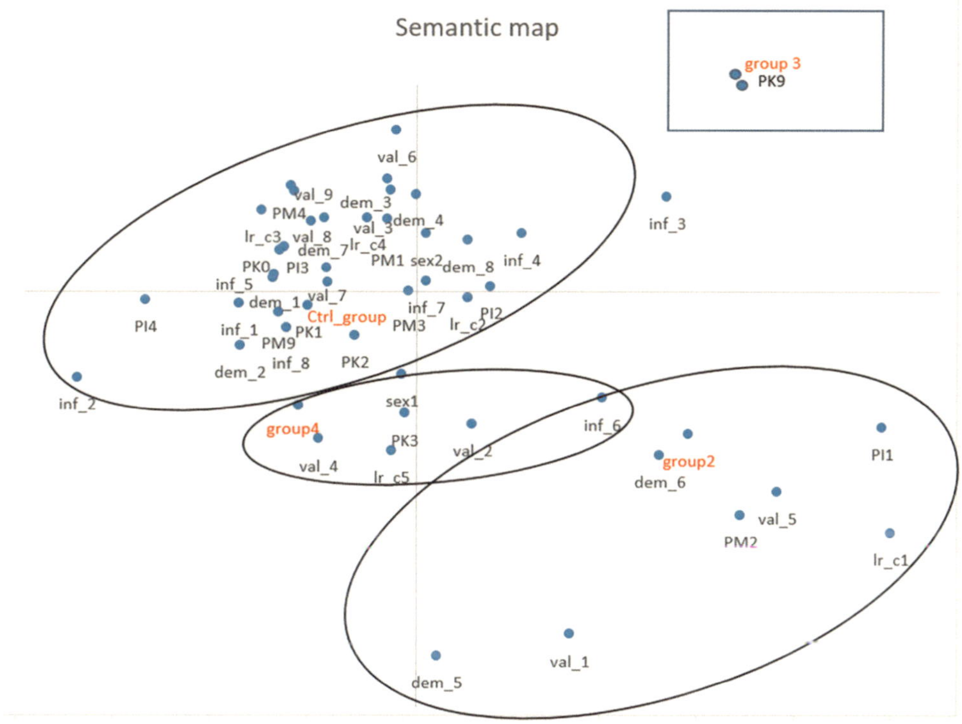

Figure 6.
The semantic map, visualizing in a Cartesian field (x,y) the categories of all variables positioned according to their coordinates from AFC.

interest, left, getting informed by newspapers and social media, expressivists choosing collective ways of mobilization.

The second axis depicts the antithesis between group 3 (online students of the civic education course) who are connected to the online information about politics, in contrast to the in-class students of group 2 who are linked to collective ways of mobilization. Additionally, the second axis is described by the antithesis between the set val_1 (Riot, Anonymous, Army, Protest), dem_5 (Riot, Deliberation, Volunteerism, Clientelism, Rebellion, Protest) and the set val_6/val_9 (Mountain, Family/Mountain, Family, Intimacy) and dem_3/dem_4 (Ancient Greece, Representative, Deliberation /e-Democracy). This polarization is explained as the difference between the democratic and moral discourses which were detected in the analysis.

5. Conclusion

The method presented in this chapter, as applied in the example of a survey among universities and high school in Thessaloniki, follows the application of HCA and MCA (or AFC) in two steps.

The added value of the presented methodological approach lies in its competence to utilize an advanced clustering method that incorporates the dimension reduction function of correspondence analysis. Clustering in multiple stages of the analysis, produces summarized variables that can describe the overall behavior or profile of the subjects. Then these new cluster membership variables can be associated with the categories of the variables used in the clustering analysis, therefore we can associate each cluster not only with its subjects but with the categories as well. In the second step, the joint analysis of the cluster membership variables together with the rest of the variables of the analysis, produces a comprehensive clustering of all items together, associating them again with the categories of the variables. This procedure allows the researcher to have a full and comprehensive overview of the profiles of each cluster.

Moreover, correspondence analysis brings forward the inner competition of the phenomenon, extracting multiple dimensions that explain the dynamics within it. The coordinates of each object give a better understanding of the distances between them, and when analyzed again with HCA we get the final fully described clusters. The coordinates can visualize the phenomenon in a simple two-dimensional space or even of more dimensions, where the observer can comprehend in more detail the revealed inner relationships or oppositions among the subjects and the objects of the analysis.

Author details

Theodore Chadjipadelis* and Georgia Panagiotidou
Aristotle University of Thessaloniki, Greece

*Address all correspondence to: chadji@polsci.auth.gr

References

[1] Galbraith JI, Bartholomew DJ, Moustaki I, Steele F. The Analysis and Interpretation of Multivariate Data for Social Scientists. London: Chapman & Hall/CRC Statistics in the Social and Behavioral Sciences; 2002

[2] Benzècri JP. L'analyse des donnees. Tome 2: L'analyse des correspondances. Paris: Dunod; 1973

[3] Greenacre M. Biplots in Practice. Bilbao: Fundación BBVA; 2010

[4] Panagiotidou G, Chadjipadelis T. First-time voters in Greece: Views and attitudes of youth on Europe and democracy. In: Chadjipadelis T, Lausen B, Markos A, Lee TR, Montanari A, Nugent R, editors. Studies in Classification, Data Analysis and Knowledge Organization. Springer, Cham; 2020. pp. 415-429

[5] Chadjipadelis T. Parties, Candidates, Issues: The Effect of Crisis, Correspondence Analysis and Related Methods. Napoli, Italy: CARME; 2015

[6] Papadimitriou G, Florou G. Contribution of the Euclidean and chi-square metrics to determining the most ideal clustering in ascending hierarchy (in Greek). In: Annals in Honor of Professor I Liakis. Thessaloniki: University of Macedonia; 1996. pp. 546-581

[7] Karapistolis D. Software Method of Data Analysis MAD [Internet]. 2010. Available from: http://www.pylimad.gr/ [Accessed: January 25, 2022]

[8] Greenacre M. Correspondence Analysis in Practice. Boca Raton: Chapman and Hall/CRC Press; 2007

[9] Marangudakis M, Chadjipadelis T. The Greek Crisis and its Cultural Origins. New York: Palgrave-Macmillan; 2019

[10] Taylor C. Sources of the Self. Cambridge, MA: Harvard University Press; 1991